D0042062

THE
MOUNTAINS OF
SAINT FRANCIS

THE
MOUNTAINS OF
SAINT FRANCIS

Discovering the Geologic Events
That Shaped Our Earth

Walter Alvarez

W. W. NORTON & COMPANY
New York • London

Copyright © 2009 by Walter Alvarez

For information about permission to reproduce selections from this book,
write to Permissions, W. W. Norton & Company, Inc.,
500 Fifth Avenue, New York, NY 10110.

For information about special discounts for bulk purchases, please contact
W. W. Norton Special Sales at specialsales@wwnorton.com or 800-233-4830

Manufacturing by RR Donnelley, Harrisonburg
Book design by Helene Berinsky

Library of Congress Cataloging-in-Publication Data

Alvarez, Walter, 1940–
The mountains of Saint Francis : discovering the geologic events that
shaped our earth / Walter Alvarez.
p. cm.
Includes bibliographical references and index.
ISBN 978-0-393-06185-7 (hardcover)
1. Geology—History. I. Title.
QE11.A48 2008
551.7—dc22

2007034311

W. W. Norton & Company, Inc.
500 Fifth Avenue, New York, N.Y. 10110
www.wwnorton.com

W. W. Norton & Company Ltd.
Castle House, 75/76 Wells Street, London, W1T 3QT

1 2 3 4 5 6 7 8 9 0

This book is dedicated with much affection and gratitude
to the geologists of Italy, who for many years
have extended their welcome and friendship to me
as together we have made wonderful discoveries.

CONTENTS

PRELUDE

In which, from a mountaintop, we look out across the landscape of Italy and prepare for a journey into the deep past of the Earth. On that journey we will come to understand how landscapes emerge from the past, and we will explore the great mystery of time—the matrix in which our lives, the history of humanity, and the history of the Earth are embedded.

FROM HIGH UP on a peak called Monte Nerone, on clear, crisp autumn mornings, you can see far across the landscape of Italy. I often go there while consulting the secret archives of Earth history, written in the rocky outcrops of the Apennine Mountain Range. Stretching all around the horizon, the view carries you back further and further into the past.

To the west range after range of lower mountains recede across Umbria toward sunny Tuscany, where in the 1660s Nicolaus Steno first discovered that the Earth has a history, and learned to read that history in the rocks of the gentle Tuscan hills. Everywhere the scenery tells its own geological story—a story that connects with the human past as well.

Around to the south you can make out high, rounded Monte Subasio, where almost five centuries before Steno, Saint Francis of Assisi began a spiritual journey that would profoundly change the world, from his medieval days to our own time.

Off to the east a prominent mountain is slashed through by the wild canyon of Furlo, where seven centuries before Francis, Goths and

Byzantines struggled to control a road called the Via Flaminia, built by the ancient Romans.

And along the northern horizon stretches a wilderness of scrambled rock that has glided, mysteriously and imperceptibly, a hundred miles across Italy, hinting at the almost unthinkable wilderness of time that recedes back into our planet's past.

Earth's rocks, like the craggy outcrops on Monte Nerone, bear the record of ancient Earth history. Written in the rocks, though not by a human hand, chronicles of strange and marvelous events await us. In the rock archives we can learn of a huge but painfully slow collision between Italy and Europe that forced up the jagged barrier of the Alps. And we can read of another collision, this one blindingly fast, in which a comet or asteroid struck the Earth, a catastrophe that spelled the end of the dinosaurs. And the archives tell of volcanoes erupting where the city of Rome now stands, and of a time when the Mediterranean Sea evaporated, leaving a vast desert far below sea level.

I have had the great privilege of working for many years in places like Monte Nerone with the Italian geologists, who have uncovered a whole pageant of Earth history. Their annals stretch back hundreds of millions of years, helping us to understand the origin of their beautiful land, and to appreciate the even broader saga of historical evolution through which our entire planet has taken form.

After a day of hard work in the mountains with my Italian friends, we might find ourselves in a little, dimly lit tavern in some medieval village hidden back in a remote canyon of the Apennines, where local wine and pasta would be waiting. After dinner, sitting around the fire, we might tell stories of the adventures and discoveries of our own geology teachers, or our teachers' teachers, and perhaps join in the soft, sweet harmonies of an old Italian song from centuries past. We might take a chilly walk through the ancient heart of town, its narrow streets closed in by stone buildings that still remember the Middle Ages. And sometimes a full moon dancing through gaps in dark clouds would complete the illusion of being transported far back into the remote past.

In a place like this, human history and Earth history seem intimately

intermingled. Both are written in stone—the stones that nature has created, and those shaped and set in place by our human ancestors. Most of
us have at least some familiarity with the history of humanity, but few
people other than geologists find the history of our planet familiar. Yet this
kind of history abounds in wonderful stories of worlds that sound like the
settings of science fiction or fantasy novels but come from the real past of
the Earth. The drama of this planetary history rivals the great episodes in
human history—the Roman Empire, the Middle Ages, the Renaissance,
the age of exploration, or the two world wars. In this book I hope to make
this historical treasure more widely accessible—to bring to life both the
historical worlds that geologists have uncovered, and the geologists who
have made these discoveries.

It is not easy to tell the story of our planet's past. Earth history is
painted on a canvas so multi-textured and complicated, stretching so far
back in time, that one must pick and choose what to include, trying to
give a broad appreciation and a general flavor, because an encyclopedic
account would overwhelm anyone who tried to read or write it. So I have
chosen to focus this account of Earth history on one place—the backbone of the Italian Peninsula between Florence and Rome that is officially
known as the Umbria-Marche Apennines, although I like to call it "the
Mountains of Saint Francis."

Writing about Earth history is also tricky because reading the past
recorded in rocks is an arcane skill of little use in everyday life, so therefore is not in the intellectual toolkit of most people. It seems especially
foreign because of the immense stretches of time and the almost painfully slow rates of change in the remote past, and this calls for a different viewpoint. Our students at Berkeley enter as freshmen with no more
appreciation of geological time and rates of change than anyone else, but
after a couple of years of classwork they are fully at home in that arena. I
have tried to distill here the concepts and attitudes and information that
geology students acquire in a few years and to streamline the process of
learning to think comfortably about the deep past.

The mountains, deserts, hills, and valleys of the entire world are the
laboratory of the geologist, and they also provide the library in which we
read the history of the Earth.[1] For those of us fortunate enough to be in this
line of work, geology is an exhilarating intellectual adventure, carried out

in the open air in the company of good friends. In addition, discovering Earth history is often a true physical adventure as well, for we live and work in remote parts of the world, traveling through wild and challenging country and studying rock outcrops in beautiful and little-known landscapes.

Geologists cannot help meeting the people who live in those places, and getting to know them and their cultures is one of the fringe benefits of our science. If one has a passion for Earth history, it is easy also to become fascinated with the *human* history of the places where we work. And so in this book I cannot resist setting the geological story of Italy in the context of the people who live there and the human history that lies behind their culture.

Geological history has happened on such an enormous timescale—millions or even billions of years—that it is almost impossible to comprehend. Even human history, stretching over centuries and millennia, is daunting to the minds of individual human beings, whose lives are measured in decades. To keep these timescales in balance I have tried to weave together stories at all three levels—first the geological history of Italy itself, then the human history that has unfolded there, and finally my own experiences and the adventures of the geologists I have met or heard about in the Italian mountains.

In organizing the geological history of this portion of Italy, I found it best to tell the story partly in reverse chronological order. So at the beginning we will be peeling off the layers of history and journeying progressively back in time. That has the advantage of starting with the comfortably familiar Italy of our own day and gradually unveiling the unexpected worlds that used to be.

After an introductory chapter that sets the scene in Assisi, the home of Saint Francis, we travel to Rome to explore the role of volcanoes in the very recent, though prehuman, geological past. The landscape of the Eternal City is sculpted from these very young volcanoes, and volcanoes provide a fine introduction to geology, for almost everyone has seen pictures of volcanoes and of their eruptions, and these eruptions take place in the familiar environment of the dry land on which we live our lives.

Under the volcanic rocks of Rome lie older deposits—sedimentary rocks that tell of a time when Italy lay beneath the sea. These deposits raise the question of how we can determine the ages of rocks, for without knowing the ages, we have no hope of understanding the history of the Earth. To trace the three-century detective saga in which geologists have discovered ways to date ancient rocks, we travel first to Siena, in Tuscany, and then east to Gubbio, on the edge of the Mountains of Saint Francis.

Finally, on our journey back in time, we confront the Apennine mountain range itself, in all its mystery, complexity, and majestic beauty. For this part of the story it works better to reverse our chronological viewpoint and follow the Apennines through time as they developed from a submerged marine sea floor into the landscape of peaks and canyons we enjoy today. In doing so we will explore some of the most surprising and profound discoveries of the geologists.

A few explanations are in order: After debating whether to use the metric units native to Italian geology or the English units native to most readers, I ended up using whichever seemed more natural in a particular situation. Since most measurements in this book give just a general sense of size, there is no need to make careful conversions between English and metric units. At this descriptive level a mile is about a kilometer and a half, a meter is about a yard, and a centimeter is close to half an inch.

Geologists familiar with Italy will recognize that I have focused on older pioneering studies and have not tried to be completely up-to-date with the current literature. This is because new, interesting papers appear so frequently that any attempt to honor them would quickly become dated. On the other hand, the great discovery papers will always remain classics. Unfortunately our scientific training tends to emphasize the most recent work and to pass over the older papers that reported the discoveries on which everything subsequent has been built. In this book I hope to counter that tendency by honoring the great discoveries that sadly are less and less remembered.

Geologists whose native language is English may be surprised at the lack of emphasis on familiar names from the iconography we all learn—James Hutton, William Smith, Charles Lyell, Charles Darwin, and all the other founders of British geology. This was done on purpose, for I hope this book will serve geologists as an antidote to an Anglophone viewpoint that ignores many of our worthy scientific forebears from other countries.

PART I

———— ✸ ————

Assisi

In which we visit the little medieval city of Saint Francis and encounter the three great questions of the book: How can we discover the strange early worlds that once existed on Earth? How can we learn the ages of rocks, and of the events in Earth history that the rocks record? And how have the mountains and valleys of Earth's present landscape come to be?

I

ASSISI IN THE WINTER

I REMEMBER the bitter cold on the day after Christmas in 1970. As Milly and I drove north from Rome, into the Apennines, snow lay in patches on the fields, and caps of white crowned the peaks. Beneath gray skies we emerged from an icy, mountainous tract and entered the wide valley of the upper Tiber River, flanked by frozen hills on either side.

Milly and I had been married only five years, but already we had lived in South America, in Holland, and in Libya, for geology is a wonderful path to adventure. Now we were living in an almost unchanged medieval village north of Rome, and Christmastime offered the chance to explore a new part of Italy.

Earth history fascinated me. As a young geologist trained to think about the evolution of mountain landscapes, I could not help wondering why the Tiber was flowing in this wide valley between low hills. So far from the sea, I would have expected a fresh, vigorous river cutting a deep canyon into the bedrock. What historical events, way back in the planet's past, might have produced this unexpected landscape?[1]

The road wandered through the frozen valley, and then, ahead of us, a much higher mountain came into view—massive, round-topped Monte Subasio, covered with snow, dominating the landscape around it.

It seemed out of place, like a king among commoners. What kind of history could have produced this huge, isolated mountain?

On a shoulder of Monte Subasio, just above the valley floor, stood a little medieval city. From a prominent church high on the left, at the near end of the town, the buildings descended to the right in a graceful sweeping curve, punctuated by other, smaller churches. The effect was dramatic and elegant. It was our first sight of Assisi.

Milly spent her childhood in Virginia, but I grew up in California, looking across San Francisco Bay toward the city named for Saint Francis of Assisi. All my life I had used the Spanish version of the saint's name, but I knew little about Francis himself. Assisi would give us an opportunity to find out—to join human history with that of the Earth.

Assisi at Christmastime

Assisi looked almost deserted—anyone who could stay inside was keeping warm by a fire. The road climbed up past old stone buildings, gray in the winter light, entered the town, and then opened out into a piazza dominated by the great medieval church we had seen from the valley below. This was the Basilica of San Francesco, the main thing we had come to see.

The Basilica of Saint Francis actually consists of two churches, one on top of the other. From the intimate and ornately decorated lower church, a staircase leads down to the crypt that contains the tomb of Saint Francis. The austere upper church was built in the Gothic style, its walls decorated with a series of frescoes by Giotto, depicting the life of Francis with a realism that broke with the symbolic conventions of medieval painting and foreshadowed the coming Renaissance. The Giotto frescoes in Assisi mark one of the great turning points in art history.

I remember the numbing chill in the completely empty basilica, where Milly and I were all alone for an hour or two, as the paintings of Giotto taught us about who Saint Francis had been, what he had believed, and what he had done. We saw Francis thrusting his rich clothes back at his father, choosing to embrace poverty. We saw him giving his cloak to an impoverished knight. We saw Pope Innocent III dreaming that this devout beggar, through his humility and poverty, might somehow save

The Basilica of San Francesco in winter. Although I did not take a picture at the time, this image, by the noted Italian photographer Elio Ciol, perfectly captures the atmosphere.

the Roman Church. Milly found herself wondering how comfortable Francis—*Il Poverello*, the little poor man—would have been with the huge and elaborate basilica built to honor his memory.

As we wandered through the two churches and the adjoining cloister, a geologist like me could not help noticing that the building was constructed of a beautiful limestone. Some blocks were white, some a startling pink color, and many were pink with dramatic streaks of white. Scattered through the limestone were tiny specks that I later learned were microscopic fossils. I could not possibly have imagined that those pink-

and-white rocks, exposed in mountainsides and quarries all over this part of Italy, would, over the next decade, lead my friends and me to a pair of remarkable discoveries about the history of the Earth.

One of those discoveries would allow geologists to date the motions of continents and would thus play a role in the plate-tectonic revolution that was fundamentally altering our understanding of the Earth. The second discovery would overturn the view that all changes in the Earth's past have unfolded slowly and gradually—the most strongly held belief of geologists and paleontologists about the nature of Earth history.[2] I could not then have imagined how well I would come to know those pink-and-white rocks!

Finally we were compelled to leave the church and its frescoes and to seek warmth. Partway across the deserted town, through a stone doorway and down a couple of stairs, we found an unpretentious little trattoria that offered a noontime meal. It was cold in there as well, until the proprietor brought out a bucket of sand with live embers on top and put it beneath the table. In that delightful warmth we enjoyed a plate of spaghetti.

Evening in Assisi at Christmastime was magical. Saint Francis is said to have been the originator of the *presepio*, or Christmas crèche. Everywhere in the town there were Nativity scenes of every imaginable kind, illuminated by candles or lanterns, and everywhere we could hear the strains of a favorite Italian Christmas carol:

> *Tu scendi dalle stelle, o Re del cielo . . .*
>   "Down from the stars you come, O King of Heaven . . ."

And in a little hotel called Albergo Sole, there was a blazing fire on the hearth, where an old lady named Peppona was cooking the flat bread called *crescia*, which mingled its aroma with that of a delicious sauce for the pasta. It was good to be inside with Milly, warm firelight illuminating the stone arches of centuries past, as a gentle snow fell outside the window.

History Preserved

All this I remember. Some of the memories are vivid, some are hazier, but they are my direct, personal mental record of an episode in the past—the most direct way we have of knowing history. Those memories of Assisi come from all five senses—the sight of the basilica with its pink-and-white limestone and the Giotto frescoes, the sound of the Christmas carol, the smells and tastes of dinner in the hotel, and the feel of the cold in the church and the warm coals under the table.

Ideas form part of the memory as well, for we were not only gradually becoming aware of geological puzzles like the flat valleys within the Apennines but learning about who San Francesco had been and realizing the impact of his thoughts and beliefs on the medieval and modern worlds.

Memories like these are the first and most immediate of the ways we know the past. Memories go back only a few decades at most and are frail and unreliable, but there are other ways to bring history to life. Books and documents carry us back past the memories of the living and make possible a kind of one-way conversation across time with writers of the past.[3] Books and documents are the main resource of scholars of human history, extending back about five thousand years. However, they help geologists only with the extreme recent end of Earth history. They make it possible, for example, to document slight variations in climate, such as the medieval warm period that allowed the Norse to farm in Greenland, and the subsequent "little ice age" that produced the frozen canals of Holland painted by Pieter Brueghel the Elder in the sixteenth century.

Another great treasury of information about the past is found in buildings, both those still standing and those now fallen into ruin and covered with younger deposits. Along with ancient buildings are other human artifacts, preserved because they have lain buried in the ground. The archaeologists who study this buried historical treasure find that things made of wood, leather, and metal have usually decayed beyond recognition, but stone tools, statuary, and buildings may be beautifully preserved, and so we realize that history can be written in rocks.

And the history recorded in rocks goes far beyond the human artifacts found in archaeological excavations. To approach the history of the *Earth* we need to make a leap to a less familiar idea—the concept that the history of our planet before humans existed is written in the rocks we see all around us. Rocks are the primary archive of Earth history because they are solids. Solids remember; liquids and gases forget.

Most people who go for a walk in the countryside find their interest drawn to trees, flowers, birds, and animals—things that move and grow and change. Rocks are less interesting to most people, except as part of the scenery, because they just lie there and don't change. But it is the unchanging, or slowly changing, character of rocks that makes them such good recorders of the Earth's distant past and explains their fascination for geologists.

When we begin to appreciate the idea of rocks as recorders of the truly ancient history of the Earth and start to learn what happened in that history, we experience a dizzying but exhilarating expansion of our appreciation of time. It's like taking off in an airplane, rapidly climbing to cruising altitude, and suddenly seeing our narrow surroundings unfold into a vast and intricate landscape—in this case the landscape of history.[4]

Here we come face-to-face with perhaps the greatest contribution geologists have made to human thought—the discovery of deep time. This jump in timescale is the most challenging step we will have to take, but also the most rewarding, for it opens sweeping panoramas of history. It will bring the recognition of astonishing events that have changed the course history was following and that have led, by the most remotely improbable pathways, to the individual people we are today and the kind of world that is our home. To me it is sad that these vistas and panoramas are known to so few, mostly to geologists and paleontologists. They deserve wider familiarity, and that is my goal in this book.

As we explore the past of Italy, we will find an exceptional richness of history recorded in stone. From the medieval Basilica of Saint Francis in Assisi, to the ancient buildings of classical Rome, to the volcanic ash that fell on the site where Rome would be built, to the sediments deposited on the sea floor that were to become the Mountains of Saint Francis, there may be no better place than Italy to read the historical archives written in rocks.

The Fascination of Quarries

It was three years after our first trip to Assisi at Christmastime. Milly and I had returned to the city of Saint Francis on a geological reconnaissance trip with our friends Bill and Marcia Lowrie. Bill, born in Scotland, is a paleomagnetist—a scientist who can read the fossil compasses preserved in rocks. These fossil compasses can tell us where continents were located on the Earth eons ago, and how they were aligned relative to north. We were there to collect samples of Apennine limestones, and we hoped that if we could read the fossil compasses in the rock samples, they would help us understand how the mountains had come into being, so that we could draw accurate maps of Italy in very ancient times.

To collect the samples we needed, we headed out to some old quarries around on the hillside behind Assisi, along a road that leads out a gate in the town wall, close to the Basilica of Saint Francis. There, on a gloomy October day, we met a quarryman named Bruno Bovi, who welcomed us and showed us the age-old techniques of Italian stonecutters.

Visiting a quarry may not be everyone's idea of an agreeable way to spend an afternoon. There are even those who consider the Assisi quarries ugly scars on a beautiful wooded Apennine mountainside. But for geologists quarries are wonderful places, allowing us to see down through the trees and pastures that may be pleasing to the eye but obscure the profound stories that rocks have to tell. In a quarry we are privileged to look into the heart of the mountains and open up the record of the Earth's past. For me, finding a new quarry can be an emotional experience akin to entering the granite portal of the university library at Berkeley. In a library and in a quarry we confront the archives of history.

For centuries quarrymen have been extracting a beautiful limestone from the Assisi quarries. Italian geologists call it the *Scaglia* (meaning scale or flake, and pronounced *Scáhl-yah)* because it can be shaped into building stones by striking off smooth scales or flakes. The *Scaglia* limestone of the Assisi quarries is pink and streaked with white. Elsewhere in the Mountains of Saint Francis the *Scaglia* is either pink or white, but not streaked, as it is at Assisi. The streaking is the same as we saw in the building stones of the Basilica of Saint Francis. Were these the very quarries used to build the great monument to Francis? Almost surely so, because this is the closest place to the basilica where you can get good building stone, and the coloring is unique.

Later we learned the dramatic story of the construction of the basilica. Milly was far from the first to imagine the distaste Francis himself

One of the old quarries behind Assisi, where Brother Elias probably got the pink-and-white limestone to build the Basilica of Saint Francis. Milly Alvarez and Marcia and Bill Lowrie are barely visible at the bottom center, collecting samples on this gray day in October 1973.

would have felt for the great church built to be his sepulcher and monument. Saint Francis had insisted that the brothers of his order embrace poverty, own nothing, and lead lives of humility and deprivation. After his death Brother Elias, one of Francis's closest companions, embarked on the building of the great basilica to honor the memory of the founder of the order, a project that began the alienation of those Franciscan brothers who wanted to follow strictly Francis's teachings about poverty. Brother Elias became one of the most divisive and, probably unfairly, most reviled figures in the history of the Franciscan order, but the basilica he built is an architectural masterpiece of the Western world.[5]

Three Grand Questions about the Earth's Past

What can you see in the quarries of Assisi? Let us concentrate on three observations that set before us the main questions that will lead us into our journey through deep time.

Let us begin by examining the pink-and-white *Scaglia* limestone very carefully, using a little hand lens—the folding magnifying glass that all geologists carry. When we look closely, we see that there are tiny fossils no bigger than a pinhead, scattered all through the rock. I imagine that Brother Elias must have noticed them during the construction of the basilica—anyone looking closely will see them as little dark specks, and to a nearsighted person the largest of them are just barely recognizable as tiny shells. Even if he saw them, however, he could not have known what they were and probably would not have been interested, but to us they tell a remarkable story. These microfossils are the shells of single-celled organisms called Foraminifera, which live only in the ocean.

The simple observation that there are marine fossils in the *Scaglia* tells us that at one time, far back in the past, the sea covered the place where Assisi now stands. And that brings into focus one of the great discoveries of geology—that the surface of the Earth has been a very different place in ages past than it is today.

So our first question is this: How can we learn about the strange early worlds that once existed on Earth? The answer lies in the rocks, because rocks make up the archives of Earth history. To see how geologists reconstruct the ancient worlds of Earth, we will journey to Rome in part 2 of the book and see how volcanoes and rivers built up and sculpted the landscape of the Eternal City.

The second big observation also concerns the microfossils made by Foraminifera. If we carefully study the Foraminifera from many layers of

Scaglia limestone in quarries in and near Assisi, we will see that their shapes change as we go upward in the sequence of beds. This is called "faunal succession," and it provides the observational evidence for evolution—the change in species of organisms through time. Faunal succession was the first great method developed by geologists for determining the age of rocks, and today there are several other important tools for dating rocks.

So our second question is this: How can we determine the ages of rocks, and of the events in Earth history that the rocks record? On a trip to Siena and Gubbio in part 3 of this book we will learn how geologists gradually developed the tools needed for finding the ages of rocks, and become familiar with the great geological timescale that ties together everything that has happened in the history of our planet.

Finally, there is a third important thing to see in the quarries of Assisi. The *Scaglia* limestone occurs in the parallel layers that geologists call beds, which must originally have been horizontal.[6] Now the beds are tilted about ten degrees down from the horizontal, as you can see in the lower half of the picture on page 10. In other places in the Mountains of Saint Francis the limestone beds are tilted steeply, and some are actually vertical or even overturned.

It is clear that something remarkable must have happened in order to take all these beds of rock and tilt them away from their original orientation. Just think about how much work you would have to do to lift heavy rocks even a few feet and put them on top of a stone wall, and you begin to sense what strong forces must have affected the *Scaglia* limestones and how much energy it must have taken to tilt all the limestone now seen on the mountainsides. That makes us realize that the Assisi quarries bear witness to a great event in Earth history—the deformation of the Earth's crust that produced the Apennine Mountains.

So our third question is this: How have the mountains and valleys of Earth's present landscape come to be? Part 4 of the book will take us on a leisurely exploration of the Apennines, where we will learn of the very surprising ways in which a great mountain range has been constructed.

With these three grand questions to guide us, we are ready to embark on a journey into the little-known worlds of the past, into the trackless expanses of deep time, and into an intimate familiarity with the secrets of a beautiful and fascinating mountain range. But before we depart, Assisi has one more surprise for us.

Rainbow

The quarries of Assisi were one of the first places Bill Lowrie and I collected the samples we hoped would let us determine the ancient locations and orientations of Italy. Unfortunately we were mistaken.

Our idea did not work, but in the process of finding out that we were wrong, we serendipitously discovered something much, much better. We came to realize that the *Scaglia* limestone is a wonderful recorder of Earth history. In that record we were to learn things no one had ever known. We found we could date the very slow spreading of oceans as they get wider and wider, and we discovered the key to understanding the extinction of *Tyrannosaurus rex* and the other dinosaurs. What we found in the Apennines would eventually lead to the discovery of the largest impact crater on Earth, in faraway Mexico. Sometimes in science it is better to be lucky than to be foresighted!

Two days after our first work in the Assisi quarries, on a darkly overcast and hazy autumn morning in October 1973, Milly and I stood with Bill and Marcia by a balustrade in the little piazza in front of the Church of Santa Chiara, just down the street from the Albergo Sole. Saint Clare was a young woman of Assisi who was inspired by Francis to found her own order of women, following the Franciscan ideals, and her church lies at the other end of the town, facing toward the Basilica of Saint Francis.

The Piazza of Santa Chiara afforded a view over the flat valley below the town. Down there, among a gathering of small houses, almost indistinguishable in the poor light and the haze, a single dome stood out. It was the Basilica of Santa Maria degli Angeli, a cathedral that honors and protects a tiny church that Saint Francis rebuilt with his own hands when he first abandoned his parents to follow his spiritual quest. As we watched, the shifting clouds allowed some rays of sunlight to break through, and a perfect rainbow took form, its end resting precisely on the dome of the basilica.

Rainbow over the Basilica of Santa Maria degli Angeli, seen from Assisi, October 1973.

Among the many sights of great beauty we were to see in this fortunate part of Italy, the rainbow at Assisi was one of the most remarkable. But in retrospect it seems like something more—like an omen, a rainbow of promise. It was almost as if Saint Francis were saying, "Welcome to my mountains! Here you will find wonderful things." And indeed that is how it turned out.

PART II

❧

Rome

In which we visit Rome and its surroundings to address the first great question of the book—How can we learn about the strange early worlds that once existed on Earth? Here we will find that the history of the Earth is recorded in rocks, which tell us the surprising story that volcanoes built the landscape of Rome, in the very recent geological past.

2

An Invitation to Rome

THE TELEPHONE RANG, and a voice from the past triggered a flood of memories. "Hello, Walter! This is Albert—Albert Ammerman." Albert was an archaeologist I had known in Rome when we were both young postdoctoral researchers in the early 1970s. We caught up on the many years that had passed, and then he came to the point.

Albert had continued to study the archaeology of ancient Rome, and now he needed a geologist to work with him. Would I like to help him figure out what the landscape of Rome—its hills and valleys and the Tiber River—had looked like before the Romans came to live there? Could I help him recover a record of that forgotten world, lost in the obscurity of time? I carefully considered the pros and cons (for about ten seconds) and asked how soon we could start.

History in the Stones of Buildings

Albert met me at the Fiumicino airport in May 1988, and we drove into Rome. As Albert dodged his little car through the chaotic traffic in the maze of streets, I found myself reconnecting with this strange and wonderful city that has been the center of the world, in one sense or another, for more than two thousand years. Here the great roads of the Roman Empire converged,

17

and here medieval pilgrims streamed in from all over Europe. From here the Roman Church still spreads its network of faith over the entire globe.

As we drove I could see the human version of history recorded in stone. We passed buildings of all ages, hinting at the many stages in Rome's complicated history. On the outskirts of the city we saw monotonous, undistinguished modern apartment houses and shops, reflecting Italy's gradual recovery to prosperity after the destruction of World War II. We drove past a cluster of white, monumental 1930s-looking buildings at Mussolini's EUR administrative center, hinting at the Fascist pride and arrogance that had lured the country into that war. As we entered the city walls, the odd little pyramid of Caius Cestius Epulo and a glimpse of the Circus Maximus reminded us that this had once been the center of the Roman Empire.

Medieval and Renaissance buildings, many still occupied, some ruined, caught the eye with their gracefully aging stone walls, rounded and decaying, and their tile roofs, softened by clumps of the grasses and small plants that find footholds among the ancient stones. Walls once painted bright red-brown had slowly faded to that gentle ocher color that must remain forever in the mind's eye of anyone who has known and loved Rome.

A kindly process of aging seemed to be at work, like the weathering of rocks in the mountains. Fresh, crisp new buildings gradually soften and become permeated with an air of romance. Watching the mosaic of architectural styles passing by, I sensed again, after many years, the character of this unique city where so much history has left its record in buildings and monuments. Living with all that history, the Romans have long had a rather cavalier attitude toward the ruins of the past amid which they build their lives.[1] Ancient monuments have served them for centuries as foundations for new buildings and as sources of recycled building material.

It all made sense to me as a geologist. In the cityscape of Rome one can see, just as a geologist can see in the landscapes of nature, a series of great cycles in which times of building and construction alternate with episodes of collapse and decay. Albert and I had come here to tease out evidence for cycles of building and destruction much older than the Italian Rome of the nineteenth and twentieth centuries.

We did not know it yet, but our quest would lead us to scientific adventure on the most sacred hill of ancient Rome, to an appreciation of the role of volcanoes in molding the landscape of the city, to an understanding of the startling history of the Tiber River, and even more strangely, to a connection with the great sheets of glacial ice that once covered Canada and Scandinavia.

Maps and Charts of History

In the library of the American Academy, a center for scholars doing cultural and historical research in Rome, we pored over maps and books by earlier generations of scholars who had tried to understand the configuration of the city of the ancient Romans, buried beneath medieval and modern Rome.[2] The layout of the city began to make sense.

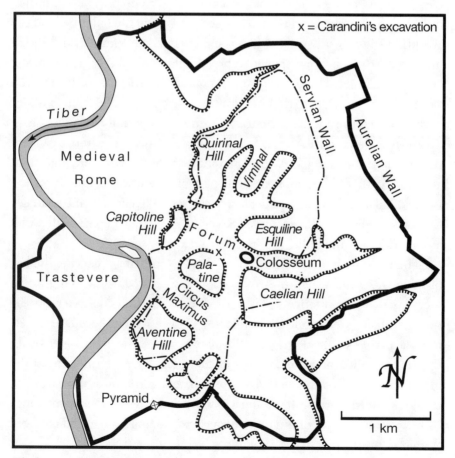

The historical center of Rome, lying within the Aurelian Wall.

The first thing that catches your eye on a map is a great construction of stone—the Aurelian Wall that surrounds the old part of the city, enfolding a vaguely circular area. This defensive line, built under the emperor Aurelian in A.D. 271 to 275, offers a physical testament to Rome's decline from greatness. At the height of the empire no enemy could even have imagined attacking this mightiest of cities. By Aurelian's time, however, Rome's power had diminished to the point where it was subject to incursions by Germanic barbarians, and the walls became necessary. Enclosing what was the urban area in the time of Aurelian, the walls encompass a small area west of the river that the Romans call Trastevere—across the Tiber—and a much larger area on the east bank.

Within that large main area, the medieval heart of Rome—a low-lying zone along the river—looks up eastward to a complicated pattern of hills. Three of them stand isolated—the Capitoline, the Palatine, and the Aventine hills—while others, like the Quirinal, the Viminal, the Esquiline, and the Caelian are spurs of a broad upland plateau that lies mostly beyond the walls. These must be the famous Seven Hills of Rome, you would think, but Albert told me that the "seven" were identified differently in different times, and hence the term doesn't really mean much.[3]

In the library with Albert, I began to see historical patterns. Within the circuit of the Aurelian Wall, different parts of the city were important at different times. The most ancient part of Rome was centered on the isolated Palatine Hill, a good defensive site when Rome was just a village. Legend says that Romulus founded Rome in 753 B.C., when he built a wall around the Palatine Hill.

The Palatine Hill lies close to the island in the Tiber, marking the last place downstream where the river was easily crossed, thus giving Rome a commercial importance in its earliest days. Settlements grew on the nearby hills as well, and the low valley between them became a meeting place that evolved into the Roman Forum. The landscape—hills, the Forum valley, and the crossing of the river—strongly influenced the placement and the layout of what was to become the most powerful city in the Western world. Albert and I wanted to understand that influence by reconstructing the original landscape and tracing how it had changed through time.

As Rome evolved from a village into a republic and then an empire, the city expanded. An early outline is recorded by the Servian Wall of 378 B.C., a small remnant of which still stands in front of the central train station. And then, after centuries in which no wall was needed, the emperor Aurelian built the wall of A.D. 271–275—a massive structure that is almost entirely preserved. The eastern part of the city inside the Aurelian Wall is on the plateau, uphill from the water of the Tiber. The growth of the city into that area was made possible by the construction, over a five-hundred-year period, of eleven great stone aqueducts. These structures, testifying to the engineering skill of the Romans, marched across the plains outside Rome on long lines of tall arches. The good water they brought to Rome from sources miles away made possible the great public baths named for emperors—Titus, Trajan, Diocletian—that mark the hills in the eastern part of the city.

As I studied the maps and documents with Albert, I found him very easy to work with. As a historian Albert knew how to gather information from written sources. In addition, as an archaeologist, he also knew how to get historical information out of the ground. Both of us found it natural to study maps intensely and to think of historical information coming from a stack of deposits—ruins or rocks—piled up in stratigraphic order from the oldest on the bottom to the youngest on top.

No single place on this planet has a complete stratigraphic sequence, in which you can see evidence for every historical phase. In deciphering the Earth's past you need to go to different parts of the world to find evidence for different episodes in geological history. In a similar way different parts of Rome record different periods of the city's history. Time passes everywhere, but it is recorded better in places where buildings are being built or sediments are being deposited. Time leaves a poorer record where buildings are being torn down and the debris removed, or where rocks are being eroded.

By compiling evidence from a variety of places, geologists and archaeologists gradually build up an overall timescale. Because our evidence occurs in piled-up material—in stratigraphic succession—we draw our timescales with older at the bottom and younger at the top. I've drawn this kind of a chart to help me understand the history of Rome. This time-

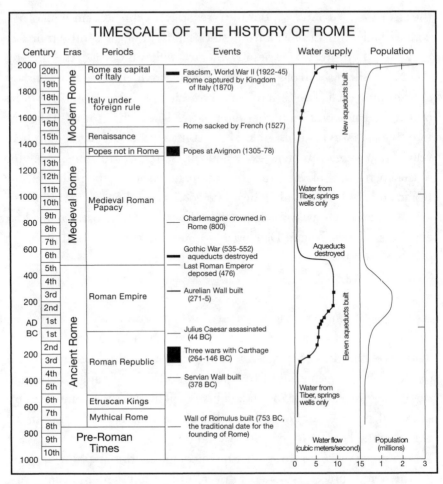

TIMESCALE OF THE HISTORY OF ROME

Century	Eras	Periods	Events	Water supply	Population
2000	Modern Rome	Rome as capital of Italy	Fascism, World War II (1922-45) Rome captured by Kingdom of Italy (1870)	New aqueducts built	
1800		Italy under foreign rule			
1600					
1400		Renaissance	Rome sacked by French (1527)		
1200	Medieval Rome	Popes not in Rome	Popes at Avignon (1305-78)		
1000		Medieval Roman Papacy		Water from Tiber, springs wells only	
800			Charlemagne crowned in Rome (800)		
600			Gothic War (535-552) aqueducts destroyed	Aqueducts destroyed	
400			Last Roman Emperor deposed (476)		
200	Ancient Rome	Roman Empire	Aurelian Wall built (271-5)	Eleven aqueducts built	
AD / BC			Julius Caesar assasinated (44 BC)		
200		Roman Republic	Three wars with Carthage (264-146 BC)		
400			Servian Wall built (378 BC)	Water from Tiber, springs wells only	
600		Etruscan Kings			
800		Mythical Rome	Wall of Romulus built (753 BC, the traditional date for the founding of Rome)		
		Pre-Roman Times		Water flow (cubic meters/second)	Population (millions)
1000				0 5 10 15	1 2 3

The history of Rome portrayed as a geological timescale. The older times are at the bottom, as they would be in sequences of geological or archaeological strata.

scale does not come from the work of a scholar of Roman history, because historians seldom use this method, but I find diagrams like this the most helpful way to visualize history.

In addition to summarizing Roman history, this diagram will help prepare for the charts of Earth history we will come to later. Let's notice a few points about it. It shows dates in years, with centuries marked, but it also breaks up Roman history into several distinct phases, like "Ancient Rome" and "Renaissance." Although a historian would not do this, I have labeled the broader phases "eras" and the more detailed ones "periods," because geologists use this system in subdividing Earth history. Another column shows a selection of major "events" in Roman history, paralleling the focus that geologists place on critical events in the Earth's past. Finally I have added a couple of columns with a plot of the water supply available to the city, increasing or decreasing as aqueducts were built and destroyed, and a plot of the population of the city, which seems largely to reflect the availability of water.[4] Geologists always look for sets of quantitative data that can be plotted on time charts to help us understand the cycles and trends in Earth history.

The Roman Forum

To begin our search for the ancient landscape of Rome, Albert took me to the Forum. The Roman Forum is the great archaeological focus of the city, where there are always excavations going on and where the ancient city is most dramatically present, almost free of medieval and modern encrustations. The Forum is a little valley, a low spot at the foot of the Capitoline and Palatine hills where the earliest Romans built their settlements. In that intermediate low space the early Romans living on the hills would meet for commerce and civic affairs, and there they constructed public buildings—their law courts, monuments, triumphal arches, and temples.

In this low area the accumulations of the centuries have built up dirt, rubbish, silt from floods of the Tiber, old pavements, and the debris of earlier buildings, demolished accidentally or on purpose. Gradually the Forum valley has filled up with layers recording the history of Rome, making it a perfect analog to the much more ancient Earth history recorded in the sedimentary layers that geologists study.[5] Beginning at the present-day surface in the Forum, an archaeologist can excavate downward through the remains of the Roman Empire, the preceding Roman Republic, the time when Rome was dominated by Etruscan kings, and with luck perhaps even back to the very founding of Rome. Below that is the pre-Roman land surface—the natural landscape before the Romans first arrived, the landscape we wanted to understand.

Albert knew all the archaeologists working in the Forum. We started at the base of the Palatine Hill, where a team was at work under the direction of his friend Professor Andrea Carandini of the University of Rome. Carandini was seeking evidence of that earliest semimythical time when Rome was founded by Romulus. So he chose to dig on the edge of the Forum valley at the foot of the Palatine Hill, where the archaeological deposits were not so thick and the bedrock of the pre-Roman landscape was not so deep and inaccessible as it would be in the middle of the Forum valley. Nicola Terrenato, a graduate student working on his thesis with Carandini and Albert, showed us around the excavations, and we particularly examined the few places where the diggers had uncovered the prehuman landscape. The bedrock they had found looked familiar to me from my work years earlier in the hilly country north of Rome, where the bedrock is made of the eruption debris of geologically very young volcanoes.

It may at first seem strange to associate volcanoes with Rome. There are no great volcanic peaks near the city, like Ixtacihuatl and Popocatepetl dominating Mexico City, Vesuvius threatening Naples, or Mount Rainier looming over Seattle. There are no plumes of ash and steam or clusters of earthquakes as ominous warnings. There are no great fields of hard black lava like those of the Hawaiian Islands. But this is volcanic country nonetheless.

Rome is flanked on the north and south by hilly areas full of the craters—some dry, some full of water—that mark the sites of volcanic explosions and collapses. The bedrock in these two ranges of hills is various kinds of tuff—soft rocks made of volcanic ash—that came out of the volcanoes in violent and dangerous eruptions. The two hilly volcanic centers flanking Rome are the Alban Hills to the south and the Sabatini volcanoes to the north. Continuing northward, they combine with two more volcanoes—Vico and Vulsini—to make up the Roman Volcanic Province. The countryside around Rome, like the city itself, bears the unmistakable stamp of the Roman volcanoes.

In the excavation Nicola showed us a volcanic tuff called the *Cappellaccio*, and I recognized it as a volcanic mudflow deposit. When volcanic ash gets soaked with water it turns into mud, and if it is on a slope, the mud can flow as a huge, gooey, moving mass. Mudflows can cause great devastation around volcanoes, as happened after the eruption of Mount Saint Helens in 1980. Mudflows may completely rearrange the landscape, damming valleys so that new lakes are formed. That would explain what we saw in another part of the excavation, where there was a bed of clay more than four feet thick in which we found fossil leaves and thin layers of crystals. Clay particles are so tiny that they can settle out only in very still water,[6] so this bed had to represent a lake rather than the flowing water of a stream. The leaves were telling us what kind of trees grew along the shore of the lake—perhaps rooted in the mudflow that had dammed the lake—and the layers of crystals were "air-fall tuffs," demonstrating that explosive eruptions had continued in the Roman Volcanic Province.

Geologists use the word "tuff" for any kind of rock made of volcanic ash. There are different kinds of tuff, deposited under different conditions, as we will see in the next chapter, and they allow the geologist to understand the environment when the tuff was deposited. Making this kind of environmental interpretation based on observation of rocks and sediments comes naturally to archaeologists and geologists like Albert and Nicola and me.

The members of Carandini's crew, under Nicola's supervision, continued to dig slowly and carefully, recording everything they found, hoping they might uncover something from the earliest history of Rome. Sure enough, about a month later they made a really big find. The remains of an ancient wall slowly emerged from the dirt. It was in just the right place to be the legendary wall of Romulus, and pottery found with the wall was of just the right age.[7]

The wall of Romulus represents more than just a marker in the growth of the city, for it was associated with a dramatic story of the founding of Rome. Writing much later, Ovid tells us that Romulus had begun to build a wall around the summit of the Palatine Hill, and left orders that no one be allowed to cross its foundations. His twin brother, Remus, unaware of this order, stepped over, and was slain by the guardian. Romulus forced himself to say, "So may all enemies cross my walls," but at the funeral he broke down in tears, saying, "Farewell, brother taken against my will."[8]

The newspapers were ecstatic about the discovery of the wall, and there was a flurry of excited news stories, adding spice to our time in the Forum. A decade later, after careful evaluation and debate, Carandini himself seemed to be completely convinced that the wall his crew had unearthed really was the wall attributed to Romulus in the ancient written sources, and he seemed irritated by continuing skepticism on the matter.[9]

Berkeley geologists Sandro Montanari and Carolina Lithgow-Bertelloni examine the "Wall of Romulus" in the Forum excavation of Prof. Andrea Carandini, June 1988.

Whether or not it is the wall of Romulus, however, it nicely illustrates the decay and loss of the rock record of older history, for the Aurelian Wall is almost complete, the older Servian Wall is fragmentary, and the still more ancient wall is known from just a few blocks of stone in one excavation.

The excavations gave us a hint of what the pre-Roman landscape was like, but the exposures of the bedrock were tiny and quickly examined. As we looked across the low-lying Forum, we realized that we would never see more than an occasional, miniscule sample of the bedrock, and only in a rare, deep excavation, because the bedrock was just buried too deeply. Clearly we would do better on one of the surrounding hills.

It was another perfect analogy to the way geologists think and work. For a geologist, lowlands are sites of deposition, where young sediments are accumulating, and if you want to study young Earth history, you go to young sedimentary basins. Mountains are places of uplift and erosion, bringing deeper, ancient rocks to the surface and exposing them, so if you are studying ancient history, you go to the mountains.[10] To find the earliest history of human Rome, we would have to go to the hills.

We checked out the Palatine Hill, whose broad, flat top had been the site of the village of Romulus, and much later was occupied by the palaces of the Roman emperors. But again on the Palatine it was hard to find out what was under the old buildings and ruins, so we ended up concentrating on the smaller Capitoline Hill, where outcrops and wells and old photographs gave us more information on the bedrock.

The Capitoline Hill

Over the years that followed, the Capitoline Hill became a source of endless fascination to me. This happens to many geologists, as they fall in love with the places they are studying, coming to know the land and its secrets ever more intimately. The Capitoline Hill was a uniquely enchanting field area. Not only did it hide geological mysteries that we slowly came to understand, but it also held a unique place in human history. In Roman times it was the head of the world—*Caput Mundi*—the site of the Temple of Jupiter, holiest place of the city that ruled the Mediterranean and much of Europe. Here defeated enemy kings were executed on the day of a triumph celebrating a Roman military victory, and traitors were thrown to their death from the high rock named for the young woman Tarpeia, the first Roman traitor. After the fading of Roman glory, this remained the emotional heart of the city, where powerful medieval families built fortresses, where Michelangelo designed his magnificent square—the Piazza del Campidoglio—and where nineteenth-century Italy celebrated its reunification with a giant monument in white marble.

It is a small hill—the Capitoline—only a third of a mile long and half that in width, roughly peanut shaped, with one end pointing north and the other southwest. It is not tall, rising only about a hundred feet above the surrounding streets, although it once stood taller, before the debris of centuries built up the street levels. It has two high points—the Arx at the north end and the Capitolium at the south—separated by a saddle where Michelangelo's Piazza del Campidoglio is located. A tiny hill, yet overwhelmingly rich in history.

On such a small hill, there is room for only a few buildings. The dominating structure is the great white marble monument the Romans call the Vittoriano, commemorating Victor Emmanuel II, the king who reunified Italy in the nineteenth century. Next to the south, crowning the summit of the Arx, is the church of Santa Maria in Aracoeli, now partly eclipsed by the great mass of the Vittoriano, but keeping alive the memories of the Middle Ages on the Capitoline. Here at the Aracoeli, Edward Gibbon conceived the idea of writing *The Decline and Fall of the Roman Empire*.[11]

Lower down, in the saddle, is Michelangelo's wonderful Renaissance composition, the Piazza del Campidoglio, framed by three palaces. The Palazzo del Senatore, in the center, is built over the massive record office of ancient Rome, the Tabularium, which looks out over the still lower area of the Forum. Atop the Capitolium summit at the south end of the hill is the Palazzo Caffarelli, built during the Renaissance over the foundations of ancient Rome's long-destroyed Temple of Jupiter, which are still partly exposed to view.

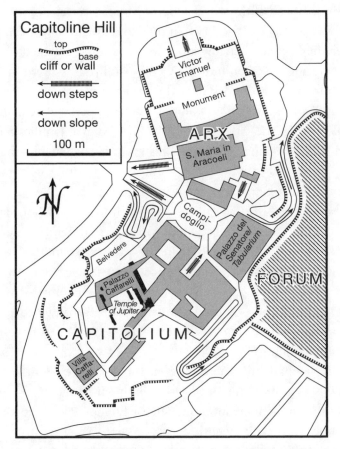

The Capitoline Hill, once the focal point of the Roman world, rises above street level on three sides, and above the archaeological zone of the Forum valley on the east.

The hill is small and easy to explore, clearly visible from its surroundings, and the buildings seem to form an elegant, well-thought-out composition. Yet, like so much else in Rome, they are simply the current occupants of a hill that has changed repeatedly and dramatically over time. To get a sense of that history, and as a preparation for finding the pre-Roman landscape hidden within the hill, let us peel back the layers of construction and see how the Capitoline Hill of today has come to be. This will be closely analogous to the way we will later peel back the layers of geological history.

It seems natural today to be able to see the Capitoline Hill rising above its surrounding, but that is the result of the most recent phase of construction—or better, of destruction. Until the 1920s most of the Capitoline Hill was surrounded by little medieval houses and churches, climbing up the slopes and obscuring the cliffs that had made this a fortified stronghold in ancient times. In those days the hill seemed to rise gradually and naturally out of picturesque neighborhoods centered on little streets like the Via Giulio Romano and the Via Tor de' Specchi.

Mussolini's Fascists came to power in the 1920s dreaming of an Italy returned to the military and political might of ancient Rome. As a symbolic step in the glorification of things Roman, an architect named Antonio Muñoz "isolated" the Capitoline Hill. Muñoz and his crews demolished the old neighborhoods that surrounded the hill, and today it rises all alone like a museum exhibit on a pedestal.[12]

Episode after episode of destruction and rebuilding stretch back into the past of the Capitoline Hill, giving it a character so much like that of Earth history. The episode that preceded Muñoz was the biggest construction event ever in the history of the hill—the building of the giant white monument to Victor Emmanuel. Celebrating the unification of Italy and designed by the architect Giuseppe Saccone, it was constructed from 1885 to 1911.[13] The upper part of the monument was previously the site of the Tower of Pope Paul III, built in the sixteenth century as the papal summer residence, and connected to the Palazzo Venezia by a covered viaduct, elevated on arches, so that the pope could pass unobserved through the old neighborhood in between. The tower, forbidding and ominous,[14] is gone now, sacrificed along with the narrow streets like the Via Giulio Romano and the secret passageway of the pope, and only a few old pho-

Via Giulio Romano, passing beneath one of the arches of the secret viaduct of Pope Paul III, at the foot of the Capitoline Hill, where the lower steps of the Victor Emmanuel II Monument are now located, in a nineteenth-century watercolor by Ettore Roessler Franz.

tographs, maps, and paintings remain to give us a nostalgic sense of what has been lost.

The past of the Capitoline Hill is an apt metaphor for the past of the Earth. History written in rocks and in ruins is swept casually away, and we read what we can find. *Sic transit gloria mundi.*

Our First Capitoline Rocks

When Muñoz cleared away the old houses on the southeast side of the Capitoline Hill, the history they represented was lost, but something else was gained, for some of those houses had been built right up against the bedrock of the hill. After the demolition was completed, the bedrock was left exposed, and those outcrops gave Albert and me our first look at rocks that are part of the deep, hidden interior of the Capitoline Hill.

They were only the beginning, however, and by the time we finished, we had worked out a detailed understanding of the complicated events that produced this central feature of Roman history and the Roman land-scape. We uncovered a story of geological cycles of destruction and con-struction inside the hill that was every bit as interesting as the cycles of human history that produced the buildings that cover the outside. This kind of complex interplay between volcanic buildup and erosional dis-mantling is a characteristic feature of landscapes on volcanoes the world over, and the way a volcanic landscape evolves depends strongly on the kind of volcanic rocks that are erupted.

The rocks we saw in those Capitoline outcrops were tuffs—various kinds of volcanic debris that have been compacted into rock. This debris had been erupted from the two volcanoes that flank Rome—the Alban Hills to the south and the Sabatini volcanoes to the north.[15]

Volcanic rocks like these were familiar to me, because I had spent most of my 1970–71 postdoctoral year in Italy working on the geology around an archaeological site called Narce, in the Treia River valley, an hour's drive north of Rome, in the heart of the Sabatini volcanic area. Tim Potter, the young English archaeologist who directed the excava-tions, had been very interested in the geology and I in the archaeology, and we had enjoyed a fine interdisciplinary collaboration.[16] In the cliffs along the Treia and its tributaries, I had come to know these volcanic rocks and how to interpret the conditions in which they had been depos-ited, and this was why Albert had asked me to work with him. So let us take a detour from our study of Rome and visit the Treia Valley, in order to learn about the kind of geology that characterizes volcanoes the world over. Then we will be in a position to understand the remarkable stories written in the bedrock of Rome.

3

WITNESS TO THE VOLCANIC
FIRES OF ROME

To GET TO the Treia Valley, let us drive north from Rome along the modern road that follows an ancient Roman road called the Via Cassia. After crossing the Tiber on the outskirts of Rome, the Cassia gradually climbs the long, gentle slope that rises northward to the volcanoes of the Sabatini area. The Cassia reaches a crest where big road cuts expose volcanic debris and then descends into a volcanic crater, one of four craters aligned east–west. The road climbs up to the other rim of the crater, past more volcanic debris in road cuts, and then descends gently down the northern flank of the old volcano. After a short distance we leave the Cassia, turning right at a sign pointing down a narrow road to Mazzano Romano.

A Gallery of Volcanic Rocks

Following this little road we drive through a gentle countryside called Pian delle Rose—Plain of the Roses—where rolling fields are interrupted by stone walls and occasional old stone farmhouses. The character of a landscape almost always reflects the strength of the bedrock that underlies the soil, and the gentle contour of these fields tells us that the bedrock

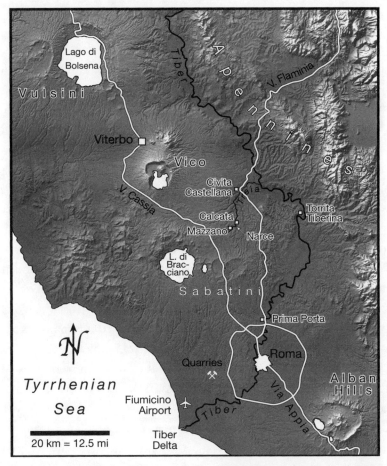

Rome and the four volcanoes. In the Vulsini and Sabatini volcanic districts, there is an east–west alignment of craters, the largest of which is a caldera, where collapse above an emptied magma chamber has formed a large depression now filled with a lake. In the Vico and Alban Hills districts, a caldera is occupied by one or two smaller lakes and a newer volcanic peak. Volcanic flows from the Sabatini and Alban Hills districts have converged on the Tiber Valley and the site of Rome.

is weak and easily eroded. Here and there outcrops of the bedrock show through, and it is the same kind of soft tuff we saw in Rome, made of tiny fragments of volcanic ash.[1] The ash particles are barely cemented together, and we can break up the bedrock with our bare hands. Bedrock as soft as this cannot stand up in steep slopes.

This is a good place to point out that geologists recognize three main classes of rock. Igneous rocks were once molten, and they either solidified at depth, like granites, or were erupted from volcanoes. Sedimentary rocks are accumulations of sediment—of grains of various sizes and origins, including pebbles, sand grains, clay particles, and fossils or fossil debris. Metamorphic rocks have been buried deeply and heated up, and chemical reactions and deformation have changed them into completely new forms. Volcanic tuffs are a kind of hybrid, because their grains are of volcanic origin but were deposited as sediment.

On the left the skyline is a rampart of low cliffs. Walking over to the cliffs, we find a hard, black rock that rings when it is hit with a geologist's hammer.[2] This is an old lava flow that was once a flood of oozing, molten volcanic magma that glowed orange-red in the night sky as it poured down an old valley on the flank of the Sabatini volcanoes. Fused into a single solid mass but cold and solidified now, it is tougher and more resistant to erosion than any of the soft volcanic ash that makes up the adjacent bedrock. The erosion that carried away much of the volcanic ash of the fields had less success attacking the lava, so it remains as a long, narrow, flat-topped ridge, marking the line of the ancient valley the lava flow followed.[3] This conversion of an old valley into a new ridge is called topographic inversion and is a key process in the evolution of landscapes in volcanic regions.

Lava flows are the volcanic event everyone knows about, from photographs or visits to places like Hawaii, or Mount Etna in Sicily, where they happen frequently. For the most part lava flows are fairly quiet and slow moving, and although they may destroy property, they rarely cause loss of life. Once, on Etna with a group of geologists, I had a chance to ride a lava flow. The surface was crusted over, with the crust broken into the jagged blocks geologists call by their Hawaiian name, aa, pronounced AH-ah. We walked carefully out onto the blocky surface, which was moving slowly as the molten lava beneath it flowed downhill. The crust kept cracking here and there, and down in the cracks between the shifting blocks, the incandescent lava warmed us with radiant heat. It was the most exciting very slow ride I've ever had. The only casualty was my field boots, which had to be thrown away because the soles melted. The point is that lava comes out in quiet eruptions that are usually not life threatening.

Although lava flows are the most familiar kind of volcanic eruption, lavas form only a very small part of the volcanic region around Rome. Most of the Roman volcanic rocks are tuffs—the hybrid rocks made of fine volcanic ash and coarser debris that was blasted out of a volcano in violent, explosive eruptions. These events are much more dangerous than lava flows, and the great volcanic disasters of history—Thera in the Greek islands in the Bronze Age, Pompeii and Herculaneum in A.D. 79, Krakatoa in 1883, Mount Pelée in 1902, Nevado del Ruíz in Colombia in 1985—were of this origin. Perhaps it is ironic that tough, hard lava rocks come from benign eruptions, while the much softer tuffs come from lethal ones.

To understand the landscape of the Romans, we must pay some attention to tuffs and how they form. The volcanic explosion provides the ash, but the ash is deposited as a sediment, so tuffs are considered to be volcano-sedimentary rocks. As we walk back across the fields of Pian delle Rose toward the road, we see some of these tuffs, very soft and easily eroded to make the gently rolling landscape. First we see beds of black pumice, very even in thickness, called air-fall tuffs. They record nearby volcanic explosions that blasted debris into the air—debris that fell out of the sky

Two beds of black air-fall tuff—deposits of volcanic fragments that fell from the sky after explosive eruptions in the Sabatini volcanic district. The inevitable geologist's hammer shows the scale.

The deposits of several lahars, or volcanic mudflows, are exposed as soft layers in this fresh road cut north of Rome.

and blanketed the landscape. Much of Washington and Oregon received a layer of air-fall ash after the eruption of Mount Saint Helens in 1980, and this happened again and again in the region of Rome in prehistoric times.

Another kind of volcano-sedimentary rock common here is the mud-flow tuff. After an explosive volcanic eruption much of the landscape is blanketed with fine air-fall ash. If the ash gets saturated by water—from heavy rains, or perhaps if the wall of a crater lake is broken and the water floods out—all that ash turns into a soupy mess of mud and can flow down-hill, down the slopes of the volcano and into stream valleys. Geologists use the Javanese word *lahar* for these volcanic mudflows, and they can be extremely destructive. After the eruption of Mount Saint Helens, huge lahars oozed down the valley of the Toutle River, carrying with them trees and logging trucks, and wiping out roads and bridges. Mudflow deposits are very common in the Roman Volcanic Province, and some flowed all the way to the present site of Rome.

Mudflow and air-fall tuffs are very soft, weak deposits, easily eroded by the streams that flow across them. The eroded ash may be carried all the way to the sea—by the Tiber in this case—or it may be stored, at least tem-porarily, in the river and lake deposits that geologists call fluviolacustrine sediments. Each of these has its own distinguishing characteristics, and as we walk back through the fields, we recognize that fluviolacustrine deposits are interbedded with mudflow and air-fall tuffs in a thick and complicated blanket of soft layers that have been sculpted into this gentle landscape.

We have now seen a lava flow and several kinds of soft, fragmented volcanic ash, and we are about to see the most spectacular and dramatic kind of volcanic rock in the region around Rome.

Back on the road, we finally reach the little town of Mazzano Romano. Turning right at the junction, we descend a steep lane, past houses built of a red-and-black volcanic stone. The houses seem to grow almost organically out of bedrock exposures of the same stone, into which wine cellars and workshops have been excavated. Looking at this rock more closely we see that there are fist- to head-size chunks of black, pumicelike volcanic material called scoria, scattered through a red matrix made of smaller bits of volcanic debris, all stuck tightly together. Both the scoria and the matrix are peppered with little holes, pores, and bubbles, sometimes filled with soft, white minerals, making it look as if the rock had once been full of tiny pockets of gas, which is indeed the case. This is a very distinctive rock, recognizable all over the Sabatini district, for which the Italian

A steep outcrop of the Tufo rosso a scorie nere, or "red tuff with black scorie," decorated at the top with laundry. The little white specks are vapor-phase minerals, crystallized from the volcanic gases that were mixed in with the ash as it flowed.

geologists use the descriptive if rather prosaic name *Tufo rosso a scorie nere*—the red tuff with black scoria.

What is this rock? It is not as hard and dense as lava. From the fragments it is clearly a tuff, but it is much harder and more resistant than the air-fall and mudflow tuffs. How did it form?

It took geologists a long time to understand this kind of rock and the eruptions that produce it. Part of the problem was that the eruptions responsible are rare, but a further difficulty was that those who witness them seldom survive to tell the story. Rocks like the *Tufo rosso a scorie nere* are the result of the most fearsome and lethal kind of volcanic events.

In these eruptions a mixture of volcanic fragments and volcanic gases comes out of the volcano at very high temperature—so hot the mixture glows incandescent red. The denser part of the mixture, called an ash flow, sweeps down the side of a volcano at high speed, scorching and flattening everything in its path. The more tenuous part rises to form a hot, roiling, glowing cloud. When the ash flow finally slows down and comes to rest, the rock fragments get glued together by minerals formed from the breakdown of glassy ash particles or crystallized from the hot volcanic gases.[4]

The resulting rock is called an ash-flow tuff, or an ignimbrite. Ignimbrite is an awkward word, but a very useful one; it comes from the roots *ign*, referring to fire as in "ignite" and "igneous," and *nimb*, meaning "cloud" as in "cumulonimbus," and *-ite*, a suffix indicating that this is a rock name. In some cases the ash flow may even be hot enough for the fragments to be soft and sticky so they get welded together, and in this case the rock is called a welded tuff. The *Tufo rosso a scorie nere* is an ignimbrite that in a few places was thick and hot enough to turn into a welded tuff. The ash-flow tuff is an excellent building material, hard enough to be strong but soft enough to be worked with picks and chisels. The welded tuff, even harder and stronger, was once quarried near Mazzano for fireplace mantels, doorsills, and the other parts of stone buildings that needed the most strength.

When I first began to study the geology around Mazzano, while working at Tim Potter's archaeological site at nearby Narce in the 1970s, I heard that there was a Roman geologist named Paolo Mattias who had been mapping this volcanic region. We got in touch, Paolo came to Narce to show me what he had been finding, and we have been friends ever since. Paolo was just finishing a geological map of an enormous area, cov-

ering all of the Sabatini volcanoes as well as the Vico volcano to the
north.[5] It is a large, magnificent colored map, and is still the basis for
more detailed local studies done since then. Paolo taught me the names
of the main volcanic tuffs around Mazzano, and in particular he showed
me that the *Tufo rosso a scorie nere* is a truly immense ignimbrite, erupted
from the Vico volcano, and extending from there all the way to Rome,
right through where the Sabatini volcanoes have formed.[6] Fortunately for
me, since Paolo had mapped such a vast area, there was still room to do
detailed studies in local areas like the Treia Valley. And thus it was still
possible to make nice little discoveries, like this one. . . .

Following the lane down the hill of Mazzano, past the outcrops of red-
and-black ignimbrite—mute witnesses to unthinkably violent volca-

*Old Mazzano, in the Treia Valley. At the left end is the ruin of a Renaissance church. Below
the houses on the right you can see the volcanic rock—the ash-flow tuff—of the pedestal on
which Mazzano was built.*

nic catastrophes in the recent geologic history of the region around Rome—we come at the bottom to a little piazza and stop at the Bar Falco to get a cappuccino from Signora Iolanda, who can fill us in on the town gossip. Then, walking through a sharply angled defensive gateway in the town wall, we come to Old Mazzano, a hill town dating back to about A.D. 900, built on a pedestal of volcanic rock in the canyon of the Treia River.

Milly and I lived in Old Mazzano for much of one year, in a couple of medieval rooms with no heat or plumbing, while I mapped the volcanic geology around the town. All the houses in Old Mazzano are built from, and on top of, an ash-flow tuff. At the end of town, where two of the three streets meet, there is a piazza where once a beautiful Late Renaissance church stood, the work of Giacomo da Vignola, a student and colleague of Michelangelo. The church was demolished in 1940, supposedly because it was structurally unsafe, and only a couple of nostalgic photographs remind us of what was lost.

When we lived in Mazzano, Reginaldo Colapietro kept an *osteria*—a very basic tavern—facing the piazza where the church used to be (the doorway is visible just to the left of the church in the old photograph). Reginaldo's wine cellar held a fascinating puzzle that was the key to understanding the volcanic landscape in places like Mazzano and Rome.

The Church of San Nicola di Bari in Mazzano, demolished in 1940, was built in 1563 by da Vignola, a student of Michelangelo.

The Mystery of Reginaldo's Wine Cellar

On summer evenings in the days before television, people used to sit on their front steps in the mild darkness. Someone would bring out a guitar and sing the old songs of Lazio—the region of Rome—songs that told stories about the rough, mean life in the Eternal City and the hard times in peasant villages like Mazzano, or amusingly insulting songs in the delightfully cynical dialect of Rome and its surrounding hill towns.

Sometimes I would be sent to get another flask of wine from Reginaldo. The *osteria* had rough wooden tables and benches, massive oak ceiling beams turned nearly black with age, one dim lightbulb, and a single calendar on the ancient plaster wall. The old men would sit around in Reginaldo's *osteria* playing the card game *briscola* in a half-light smelling of wine and smoke. At the back of the room Reginaldo would lead me down a stairway with worn steps cut into the ash-flow tuff, down to a dark cavern hollowed out of the rock. Down there Reginaldo had a couple of old oak barrels. He would let me test the wine, using the drinking glass he kept down there for that purpose, and I would depart with a big flask of wine to keep the party going.

Many houses in Mazzano had cellars like this, and there were obscure, half-forgotten passages linking them into a labyrinth of damp, moldy artificial caves that honeycombed the rock on which Mazzano was built. I had to explore those caverns, of course, to see the interior of the rock of Mazzano. Geologists love to be inside the rock, surrounded by the record of events in Earth's past.

These ancient wine cellars were cut into an ash-flow tuff—an ignimbrite—but they held a surprise. This was *not* the ignimbrite you see on the steep lane coming down to Old Mazzano, just a stone's throw away. It is not the red tuff with distinctive black scoria, but a different ignimbrite. Not only is it lower in elevation, but it looks completely different. The ash-flow tuff of Old Mazzano is yellow, not red, and it has no black scoria. What is this new ignimbrite, and what does it tell us about the history of the volcanic landscape? This was a delightful puzzle to solve, and to see the solution most clearly we need to go to Calcata, the next village down the Treia Valley to the north.

Prior to about 1960 there was no road from Mazzano to Calcata, just a difficult trail for a couple of miles among the cliffs, down to the Treia

Looking north at the medieval fortified hill town of Calcata, built on a pedestal of ash-flow tuff in the canyon of the Treia.

River and back up the other side. The two villages were cut off from each other by topography. Mazzano looked south to Rome and spoke a variant of Roman dialect. Calcata was linked northward to Viterbo, the medieval archenemy of Rome,[7] and used a dialect similar to that of Viterbo. The rivalry between the two villages has continued, and even intensified, since a road was engineered through the canyon to link them.

The road from Mazzano descends through the cliffs to cross the Treia River below the Etruscan-age hilltop fortress of Narce, the site of Tim Potter's excavation. As the road rounds the corner of the hill of Narce, Calcata comes into view, built on a rock pedestal like Mazzano, but on a much higher and more dramatic one. It is a breathtaking sight—perhaps the most spectacular fortified hill town in this part of Italy.

The most remote of the Treia villages, Calcata remained an almost unchanged medieval hill town into the 1970s, where farmers walked or drove oxcarts out to work in fields on the rim of the canyon, where water was carried in by hand to ancient stone houses crowded together on narrow streets full of children, and where timeless festivals could transport the visitor back to the Middle Ages.

How did the spectacular site of Calcata, a cliffbound pedestal of ignimbrite in the middle of a canyon, come to be? As I mapped the geology of the Treia Valley, I found that the rock of Calcata was a yellow ignimbrite, just like that of Reginaldo's wine cellar in Mazzano, and that there were a number of these yellow pedestals within the Treia Valley, contrasting with the walls of the canyon, which were made of the distinctive red-and-black *Tufo rosso a scorie nere*.

One day, while I was sitting on the rim of the canyon, studying my half-finished map and comparing it with the landscape, I realized that the pedestals of yellow ignimbrite seemed to line up. Maybe they had once been continuous! With a pencil I sketched in dashed lines grazing both sides of each pedestal. If the yellow ignimbrite had once been continuous, then the dashed parts would have been eroded away by the Treia River, with only the pedestals escaping the erosion.

My hypothetical connections between the pedestals curved gently back and forth, just as river valleys do. As the reconstructed yellow ignimbrite emerged on the map, I could see that it had a branching pattern, like the tributaries that join together in a river network.

Suddenly a possible solution came to me. Maybe there had once been a river—an ancestor of the modern Treia—that had cut down into the plateau made of *Tufo rosso a scorie nere*, with a set of tributaries joining up into a main valley. And then that valley system had been partly filled by the yellow ignimbrite—emplaced when an ash-flow from the Sabatini volcanoes poured northward down each of the tributary valleys, merging into the single large valley right at the point where Calcata now stands.[8]

It made lots of sense. After the partial filling by the yellow ignimbrite, the Treia River, now flowing on top of the yellow ignimbrite, would have continued to erode, gradually cutting away some parts, while other portions remained as the pedestals—perfect sites for villages like Calcata and Mazzano.

It seemed a reasonable explanation, but was it right? With mounting excitement, I realized that there was a way to test this hypothesis. If the yellow ignimbrite had flowed northward down an ancestral Treia Valley network, then the tops of the pedestals should be at elevations systematically decreasing toward the north. The topographic map showed

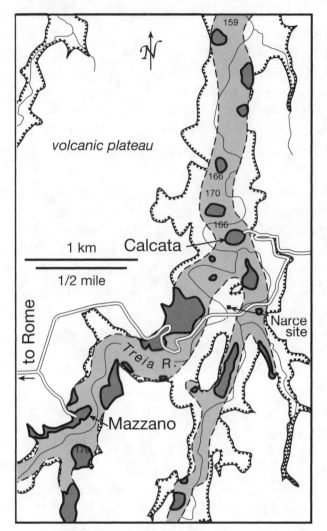

159

166

170

166

volcanic plateau

1 km

1/2 mile

Calcata

Narce site

Treia R.

Mazzano

to Rome

Map of the Treia Canyon, showing how the volcanic plateau made of the Tufo rosso a scorie nere was incised by an ancestral valley of the Treia River (light gray). The ancestral valley was filled partway up to the level of the plateau by a yellow ignimbrite, or ash-flow tuff, coming from the Sabatini volcanoes to the south, and that valley-filling ignimbrite has been eroded into remnants (dark gray), including the pedestals on which Mazzano and Calcata were built.

spot elevations in meters on the tops of some of the remnants. I penciled in the spot elevations, and sure enough, they did show a gradual slope to the north! My hypothesis had passed a critical test.[9]

It was a perfect example of scientific discovery in miniature, with the mystery, the hypothesis, the idea for a test, and the confirmation all coming together while I sat looking over the beautiful canyon that was my geological laboratory.

The Complex Landscapes of Volcanic Regions

In the Treia Valley we have been looking at details of the geology of the Roman Volcanic Province. However, it is worth backing off and trying to get a broader picture of what goes on in volcanic landscapes. I think it is fair to say that these are the most complicated and intricate landscapes we have. In most other places, where there are no active volcanoes, sediment is extracted from hills and mountains by erosion, carried downhill by rivers, and comes to rest when deposited in the ocean. The land above sea level, visible to our eyes, is dominated by erosion, while the sea floor is dominated by deposition.

Volcanic landscapes present an intermediate situation. They are the sites of a competition between the deposition of volcanic rocks and their erosion by rivers. The result is an intricate history full of intriguing little episodes like the valley-filling ignimbrite eruption and subsequent erosion that produced the rock pedestals on which Mazzano and Calcata were built. The repeated cycling between volcanic construction and erosional destruction makes the history of these landscapes particularly analogous to the history of the city of Rome, with its cycles of human building up and tearing down.

The Treia Valley has been a most instructive place. We have learned about the various kinds of volcanic rocks—lava flows and the fragmental tuffs that result from air falls, mudflows, redeposition in rivers and lakes, and catastrophic ash flows. And we have seen how the landscape evolves in volcanic regions, sometimes with the topography being inverted when harder rocks deposited in valleys stand high after softer ones are eroded away. So let us return now to Rome and use these tools to make geological sense of the Capitoline Hill.

4

The Quest for the Ancient Tiber River

In the Treia Valley, prominent rocky outcrops stand out on every side, and it is not difficult to understand the processes and history that have generated the landscape we see. In the city of Rome, on the Capitoline Hill, it is the architecture that catches the eye—the flamboyant Victor Emmanuel II Monument, the austere medieval dignity of the Church of the Aracoeli, and the Renaissance elegance of Michelangelo's piazza. The underlying rocks are covered almost everywhere by buildings, roads, gardens, and the debris of centuries. Geologists like to understand the bedrock beneath the landscape, and to know why each hill and valley is there, but in the case of the Capitoline Hill there were very few clues for us to go on.

Ancient Valleys Inside the Capitoline Hill

The first thing was to look for outcrops of bedrock. On the Capitoline Hill are just a few small, inconspicuous outcrops, but that is where Albert and I started. The best is the one labeled X on the map, on the south side of the hill next to the Piazza della Consolazione, where Muñoz's demolition crews had cleared away the medieval houses and left the bedrock exposed.[1]

Outcrop X is the best in the sense of telling us the most, for it exposes three different volcanic units, one on top of another. At the bottom, at street level, is a soft tuff the archaeologists working in the Forum call the *Cappellaccio*, which we recognized as a lahar—a mudflow tuff.[2] Above it, making a low cliff, is an orange ignimbrite called the *Tufo Lionato*, which is familiar to the Forum archaeologists. Geologists also know the *Tufo*

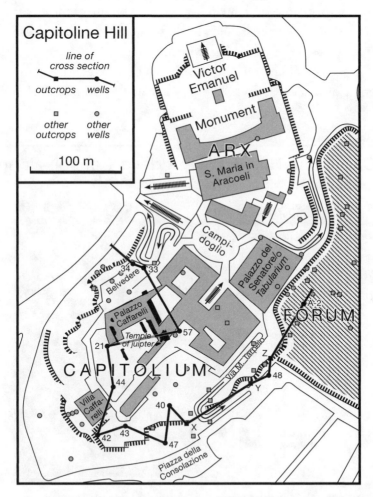

Here are the outcrops and wells on the Capitoline Hill that allowed us to find out what was inside the hill. The heavy, angled line is the location of the cross section of the hill that we drew. The cross section is as if we had an imaginary trench zigzagging around the hill and could see what rocks were in the walls of the trench.

Lionato very well and have traced its origin back to an eruption in the Alban Hills, the volcanic center just south of Rome.[3]

Then comes an interval called the fluviolacustrine unit, whose sediments tell us of a time between volcanic eruptions, when water wandered across this landscape in streams or sometimes ponded into shallow lakes. Above that is a wall that is part of the human construction on top of the hill—what Albert calls "the archaeology." Just this little sequence of layers tells us something about the prehuman history of Rome, showing that volcanic ash flows and mudflows reached the site of the city from the volcanoes that flank it. It should give us some pause when we realize that those volcanoes are probably just dormant, not extinct.

Most geological deposits are thin layers that extend for considerable horizontal distances. A layer like this is called a "formation," and it is

Outcrop X, in the cliff of the Capitoline Hill facing the Piazza della Consolazione, showing how its four geologic units are schematically portrayed in a stratigraphic column.

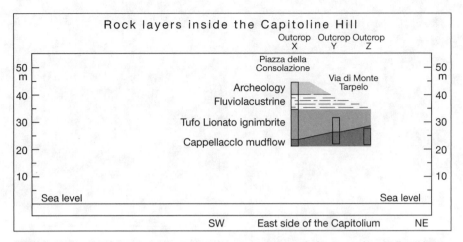

The beginning of a cross section—an imaginary trench through the Capitoline Hill. It shows three outcrops on the east side of the hill, in which we could see three geologic units, or formations, and part of the overlying human constructions—the "archaeology." Each outcrop is represented as a stratigraphic column, with the contacts between formations plotted at the appropriate elevation above sea level.

worth noting that this technical meaning of the word is different from the oddly shaped erosional remnants that are what most people have in mind when they talk about a "geological formation."[4]

Since most formations are thin layers, we can expect that at any given place the rock types will change much more abruptly vertically than horizontally. So geologists find it useful to symbolize their observations at a particular place by drawing a "stratigraphic column"—a schematic diagram of what would appear in an imaginary well drilled down through the sequence of layers. Next to the photograph of Outcrop X, facing the Piazza della Consolazione, I have drawn a stratigraphic column that will be the beginning of our study of the Capitoline Hill.

Just to the northeast Outcrops Y and Z allowed us to see that the geological contact between the *Cappellaccio* and the *Tufo Lionato* slopes gently southwestward. We can plot the three columns representing these outcrops, to see the beginning of a geological section—what you would see if you were able to cut a deep trench through the Capitoline Hill.[5]

If there had been nothing but this little bit of outcrop to go on, we could never have figured out the geology of the hill. Fortunately there was another way to get information. We could drill down through the

Above: *Albert Ammerman drilling down to bedrock with a hand auger.* Top right: *Drilling in a Renaissance courtyard in Rome.* Bottom right: *Nicola Terrenato (standing) and me, surrounded by boxes full of drill cores on the Capitoline Hill.*

cover with a hand auger until we hit bedrock, and the auger let us lift out a few chips to find out what it was. Albert seemed to enjoy this physical exercise, and over the years he has drilled lots and lots of these little hand borings all over the Forum and the Capitoline Hill. Alas, these showed us only what the very top of the bedrock was, and still told us nothing about the deep interior of the hill.

I think at that point I had pretty much given up finding out what was inside the Capitoline Hill. Not all scientific projects work out, and this looked like a hopeless one. Then out of the blue I had another call, from a very excited Albert. He told me he had just learned that the city of

Rome was carrying out a program of research drilling. The city engineers were worried that traffic vibrations in the city center might be damaging the old buildings and loosening the weak volcanic bedrock they stand on. The city government had hired a drilling crew, and they were drilling holes all over the Capitoline Hill!

Our geological research in Rome was back in action. Albert and Nicola Terrenato and I made friends with the engineers. They had been working for months, drilling holes and testing the rock cores for strength, so there were boxes of cores scattered all over the Capitoline Hill and stored away in tunnels dating back to medieval or ancient times. The engineers had no further use for the cores, so we went through box after box, identifying the rock units, noting the elevation of the top and bottom of each unit, and drawing up each well as a stratigraphic column to plot on our growing diagram.

The first few wells made it possible to understand the east side of the Capitolium—the southern crest of the hill. Two of the wells (43 and 40) were on top of the hill and showed us that the three formations in Outcrop X continue more or less horizontally, as formations usually do.

Three of the wells (47 and 48, and A-2, an older well in the Forum) were drilled down from street level, and they showed us what was under the *Cappellaccio* mudflow—the lowest unit visible in Outcrop X. Beneath

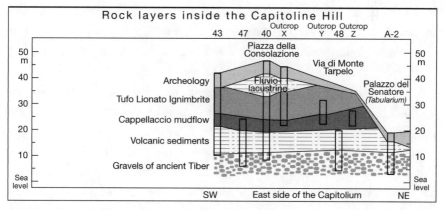

Here is the beginning of an understanding of the interior of the Capitoline Hill, with the three outcrops supplemented by five cored wells. In this southeastern part of the hill there is a series of horizontal volcanic layers resting on gravels of an old bed of the Tiber River that predates the volcanic eruptions.

the *Cappellaccio* was a formation of volcanic sediments, all full of the little crystals characteristic of volcanic ash.

Below that we came to gravels that had been deposited in the bed of a big river. It could only be an earlier bed of the Tiber, which today is just a couple of blocks to the southwest. We had expected these "Paleotiber" gravels to be there in the cores, because they are present at this elevation all around the Forum valley. Groundwater can flow easily through the gravel, and it is the source of many springs that were used by the early Romans for water supply.

We examined the gravels carefully and could not find any volcanic sand or pebbles, so this gravel must have been deposited before the Roman volcanoes began to erupt. The volcanic landscape—the Seven Hills of Rome—provides much of the character of the city, so it was striking to realize that in the very recent geological past there were no volcanic rocks here at all, and the Tiber flowed through a completely different kind of terrain.

It was great fun doing this work with a couple of archaeologists, for our ways of reading history in rocks were generally very similar, but with intriguing differences. I was comfortable with vast timescales Albert and Nicola never had to deal with, but they would always be thinking about the ways buildings and human diggings might be confusing us and leading us to make mistakes, which is rarely a concern for a geologist.

Things got more puzzling when we started looking at wells from the west side of the Capitolium. These wells went through greater thicknesses of *Tufo Lionato*, and our old friend the *Cappellaccio* was not to be seen in these cores. What was going on? When we plotted the stratigraphic columns from the western wells we could see that the *Cappellaccio* and the volcanic sediments under it were missing, and instead the *Tufo Lionato* was much thicker, going all the way down and resting on the Paleotiber gravels.

How could we account for this geometry of the volcanic units inside the Capitoline Hill? Remember that an ignimbrite is formed suddenly, catastrophically, when an ash flow roars down the slope of a volcano. That means that if we were to redraw the cross section, removing all of the *Tufo Lionato* and everything above it, we would have a profile of the landscape on the day the ignimbrite was erupted.

On that profile the pre–*Tufo Lionato* land surface (shown as a heavy

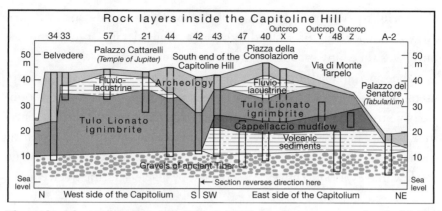

The inside of the southern half of the Capitoline Hill, based on outcrops and wells. The left half of the diagram shows clearly how the Tufo Lionato *filled a valley cut down through the* Cappellaccio *mudflow, and overflowed the* Cappellaccio *that remained.*

line in the cross section) is lower on the left side than on the right. There must have been a valley on the left. What could have eroded such a valley? The Tiber River, of course.

So when the *Tufo Lionato* ash flow swept down from the Alban Hills, it flowed into the valley of the Tiber, which had cut down through the *Cappellaccio* and the underlying volcanic sediments. This situation was reminiscent of the yellow ignimbrite of Mazzano and Calcata, which partly filled an old valley. Here in Rome, however, the *Tufo Lionato* completely filled its valley, overflowing it and covering the *Cappellaccio* as well.

So, hidden inside the Capitoline Hill was a record of two events in the history of the Earth—the erosion of a valley by the Tiber and its filling by an ash flow that came from the Alban Hills volcano.[6] We had the same thrill of discovery that a historian might have in the Vatican Archives, turning up some long-neglected document recording a forgotten episode in human history.

Volcanoes at Rome, Glaciers in Canada

Understanding the Capitoline Hill was very satisfying. It was good to be able to sit at the foot of an ancient column in the Forum, look up at the hill, and know what was under each part. It was like having an old friend whose background and character you have come to know very well. It was also useful to Albert for his study of the prehuman landscape of Rome. And yet geologically it seemed like just a single detail, of local interest only. Was there any broader significance?

There was indeed. I no longer remember when this idea emerged, for useful scientific ideas often appear from nowhere and seem obvious once we have them. At some point we realized that the Roman Volcanic Province was the key to dating the advances and retreats of the glaciers during the ice ages. At first this sounds unlikely, for Rome was too far south and too low in elevation to have been covered by glaciers during the ice ages.

At various times during the last few hundred thousand years, huge glaciers covered much of Canada, Scandinavia, and the Alps, as today they cover Antarctica and Greenland. Those glaciers waxed and waned, but they never reached Rome. The volcanic fires of prehistoric Rome and the glacial ice of Canada seem like opposites, both in terms of geography and in terms of temperature. How could they be related?

The connection is indirect but very useful. When huge ice sheets covered Canada and Scandinavia, they stored, on the land, in the form of ice, large amounts of water that would otherwise have been in the ocean. As a result sea level would fall, maybe three hundred feet or so, during a glacial episode, and rise during the warmer intervening times that we call interglacials. So the ice ages were not simply a matter of great ice sheets waxing and waning, but also a resulting history of dramatic, dynamic changes in coastlines worldwide, producing a complex geometry of eroded and subsequently filled valleys, through many cycles. A geologist looking at a shoreline thus sees it as just one frame in a long and complicated movie stretching back in time.

Like all rivers that reach the sea, the Tiber was affected by the glacial falls and rises in sea level during the Quaternary—the name geologists use for the ice age of the last 1.8 million years of Earth history. Like rivers elsewhere, the Tiber cut deep valleys when sea level was low, and filled them with sediment when sea level was high. This of course is what we were seeing inside the Capitoline Hill—a set of ancient eroded valleys, each filled up before the next valley was eroded.

The Tiber, however, is special in one very useful way. The Roman volcanoes erupted many times during the Quaternary, so the ancient valleys of the Tiber are partly filled with volcanic debris. Most rivers around the world just carry sediments eroded from the highlands drained by the river, but the fill of the Tiber Valley has this additional component of locally derived volcanic debris, which was erupted and almost immediately deposited in the valleys as they filled.

This is important because the Roman volcanic rocks are unusually rich in potassium, an element that decays radioactively to argon.[7] One of the main techniques for determining the ages of rocks in years is based on this decay of potassium to argon, so the Tiber is the ideal river for dating glacial advances and retreats. If we could date the volcanics that fill each ancient valley of the Tiber, we would have a chronology of the glacial advances and retreats in faraway Canada and, in turn, that chronology might help us understand why the ice sheets came and went through time.

It looked like a great project, but a big one—just right for a PhD thesis. Fortunately a very promising new grad student named Dan Karner arrived at Berkeley in the fall of 1992, just as Albert and I were thinking about how to do this big dating project. Dan had planned to study the Coast Ranges of California, but the prospect of working in Rome was so irresistible that he changed his mind and took on the effort to date the cuttings and fillings of the many generations of Tiber valleys that record the glacial oscillations.

We soon discovered that the best exposures of the rocks Dan needed to study are in a set of abandoned quarries between Rome and the sea, close to Leonardo da Vinci Airport at Fiumicino, where the Tiber has built its delta. Perhaps Dan should have been more cautious, for on our first visit we found that these quarries are used as the garbage dumps for all the trash from the city. It may be symbolic of the academic enterprise that the professor worked at the beautiful and sacred site of the Capitoline Hill, while the grad student worked among huge piles of rotting garbage, with tons and tons added fresh each day. And yet what Dan found in the garbage dumps was a real treasure.

We were now moving away from the area of archaeology, where Albert knew all the Italian scientists, into a more geological study, so I took Albert and Dan to meet an old friend, Renato Funiciello, professor of geology at the University of Rome. I knew that Renato and his students

Dan Karner getting to know the Roman volcanic rocks.

were interested in the volcanic rocks around Rome,[8] and it seemed that we should be in contact and perhaps develop a collaboration. It worked out perfectly, for Dan hit it off with a young researcher in Renato's group named Fabrizio Marra, who was interested in the same questions, and the two of them worked together in the field and have continued their friendship and scientific cooperation ever since. Their studies with various colleagues[9] have established in detail the timing of volcanic eruptions around Rome, and used the eruption ages to date the glacial cycles of the Quaternary.[10]

In the quarry exposures west of Rome, Dan and Fabrizio found a whole series of ancient valleys cut by the Tiber in its former incarnations, and filled not only with river sediments but also with marine sediments—because here they were closer to the sea than in Rome itself—and of course with volcanic debris. They prepared cross-section plots like our diagram of the Capitoline Hill, but much more complicated, because they could recognize many ancient valleys, eroded and then filled, over a long interval of time.

Understanding the sequence of valleys and valley fills was something that could be worked out in the field with a map, rock hammer, hand lens, camera, and collecting bag. Determining the ages of the volcanic rocks in years, with real reliability and precision, was a task requiring sophisticated laboratory techniques. Fortunately the potassium-argon dating method had been developed largely at our university, and a more advanced version called ^{40}Ar/^{39}Ar dating, has been brought to a high level here. The Berkeley Geochronology Center, a block from the campus, is one of the best dating labs in the world, and Dan and I were easily able to convince Paul Renne, the director of BGC, of the importance of dating the Roman volcanic rocks. The scientists at BGC are particularly skilled at dating very young rocks like the Roman volcanics, which are especially difficult because there has not been time for much potassium to decay.[11]

Paul introduced Dan to the laboratory techniques of ^{40}Ar/^{39}Ar dating, and together they produced a whole series of high-quality age dates on the Roman volcanics. Their dating showed that there has been a glacial advance, leading to a sea-level low stand and producing a downcut valley of the Tiber, roughly every hundred thousand years for the last six hundred thousand. The sea-level lows alternate with high stands, when large parts of the ice sheets melted away, putting more water into the oceans. This cycle had been inferred by geologists studying other kinds of evidence for the Quaternary glacial cycles,[12] but the work by Dan, Paul, and Fabrizio gave us our first direct, reliable dates on Earth's glacial rhythm during the ice ages.

After Dan finished his PhD work with Paul and me, he went on to do a postdoctoral fellowship with my old friend Rich Muller, professor of physics at Berkeley, working on understanding what drives the glacial cycles.[13] Dan has continued his work with Fabrizio Marra. They have traced the climate record of the Tiber all the way back to the beginning of the one-hundred-thousand-year glacial cycles. This work on understanding the past history of climate is a critical matter because of worldwide concern about future climate changes and their potential effect on humanity. It was very satisfying that an archaeologically motivated study of the Capitoline Hill could lead to results of real societal importance.

The Death and Rebirth of the Tiber

As creatures with life spans of only a century at most, we think of hills and mountains as permanent features. On longer timescales, however, landscapes and the rivers that sculpt them are dynamic, changing, and ephemeral. Inside the Capitoline Hill the drill cores gave evidence of a former valley of the Tiber, filled with volcanic debris, out of sight and forgotten. In the quarries between Rome and the sea we understood that there have been many cycles of valley cutting and valley filling in response to falls and rises in sea level, which in turn were due to cycles of waxing and waning glaciers in Canada and Scandinavia.

Now let us look at a still more impressive story of how landscapes can change. We shall see that there is evidence in the rock record that the Tiber was completely realigned by a volcanic eruption. Some 550,000 years ago, the Tiber was dead and buried under masses of volcanic debris, and was then reborn with a new path to the sea.[14] To examine this evidence, let us pack up and leave Rome one last time, heading north on the now-familiar Via Cassia, along the side road to Mazzano Romano, and then toward Calcata, as far as the archaeological site of Narce.

When we visited this area earlier in the book, it was to look at different kinds of volcanic tuff and to see how an old valley of the Treia River, incised down into the volcanic plateau, was partly filled by an ignimbrite flow. We saw that after this filling the Treia kept on eroding, so that remnants of the ignimbrite were left standing, to become the sites of the medieval villages of Calcata and Mazzano, and the Faliscan citadel of Narce.

The still more dramatic rerouting of the Tiber occurred at the beginning of the volcanic activity, and we can see the evidence at the bottom of the pile of volcanic rocks, where the road crosses the Treia River, at the foot of the hill of Narce. The oldest, lowest volcanic rock at Narce is a yellow ignimbrite—but much older than the valley-filling yellow ignimbrite of Mazzano and Calcata. Paolo Mattias called this older ignimbrite the Yellow Tuff of the Via Tiberina, and at the Narce site it rests on old river gravels very similar to the gravels at the base of the Capitoline Hill in Rome. As at the Capitoline, there are no volcanic grains in the gravel, so the river must have predated the volcanic eruptions. The pebbles in the gravel are made of limestone and chert, which are abundant in the Apennines, where the Tiber River has its source. And indeed this gravel at Narce seems to mark a former route of the Tiber, or what the Italian geologists call the Paleotiber.

This is not like the situation in Rome, where the old Tiber Valley and the modern Tiber are only a few city blocks apart. Here the modern Tiber is far away, fifteen or twenty kilometers east of Narce. If the Narce gravels really do represent the ancient Tiber, how did the river get shifted so drastically?

Holding that question in mind for a moment, let's look at the evidence for how the Tiberina ignimbrite affected the river that had been carrying the gravels. At the hill of Narce the ignimbrite rests directly on the gravel, so this must have been the bed of the old river. Just a stone's throw to the east of Narce, across a little tributary of the Treia, there is an old quarry where the river gravels are covered by fine-grained silt, in which there are plant remains and snail shells. This must have been the floodplain adjacent to the river channel, where fine silt was deposited from time to time when the river flooded over its bank. We get a picture of an ancient river, meandering across its floodplain, just as the modern Tiber does near Rome, at Prima Porta, for example.

Normally the course of a river is constantly shifting as its meanders migrate back and forth across its valley floor, chewing up the floodplain silt on one side and depositing a new floodplain on the other. Floodplain features are ephemeral, but at Narce the pattern of channel and floodplain was fossilized—literally cast in stone—on the day the Tiberina ignimbrite flowed into the Paleotiber Valley. It is exactly the same thing that happened at Pompeii, where a volcanic eruption entombed a Roman city, but at Narce the volcanic eruption entombed an ancient river valley.

As is the case with human history, much that happened in the Earth's past has left no record, but some events are preserved in surprising detail. We can see, for example, a remarkable record of what happened on the day the Tiberina ignimbrite poured into the old river valley. In one outcrop of the ignimbrite, at the base of the hill of Narce, there is a clear

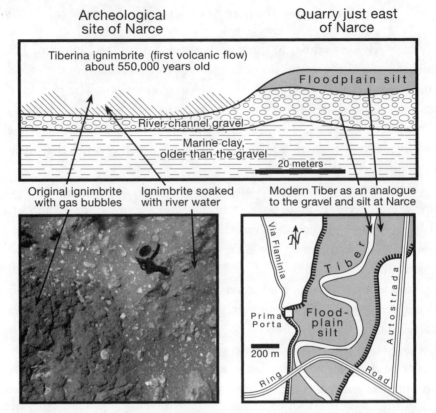

Tiberina ignimbrite (first volcanic flow)
about 550,000 years old

Floodplain silt

River-channel gravel

Marine clay,
older than the gravel

20 meters

Original ignimbrite
with gas bubbles

Ignimbrite soaked
with river water

Modern Tiber as an analogue
to the gravel and silt at Narce

Via Flaminia

N

Tiber

Autostrada

Prima
Porta

Flood-
plain
silt

200 m

Ring

Road

*Details of a disaster: The Tiberina ignimbrite, where it flowed into the ancient Tiber River
at Narce. At the top is a reconstructed cross section of the gravel bed and floodplain of
the Paleotiber based on outcrops around Narce. At the lower right is a map of the modern
Tiber, a few kilometers north of Rome, as an analog to the ancient riverbed and floodplain at
Narce. The picture in the lower left shows the lowest volcanic rock at Narce, deposited in the
bed of the Paleotiber. The rough rock still has gas bubbles, showing that it was an ignimbrite.
The smooth ash must have gotten soaked with river water and turned into mud. The steep
contact between the two probably reflects turbulence in the river water as it was invaded by
the hot volcanic ash flow. The car key shows the scale.*

transition from a volcanic tuff full of the characteristic gas bubbles showing that the ignimbrite flow was a mixture of gas and volcanic ash, to
a somewhat similar rock but without gas bubbles. The bubble-free rock
looks like a mudflow, and probably represents the part of the ignimbrite
that got soaked in the water of the river channel and turned into mud.
The steep contact between the two hints at the angry, swirling violence
where the hot ignimbrite plunged into the cold river water.

Having seen the detailed evidence that the Tiberina ignimbrite was erupted into a river channel and onto the floodplain of the river, let us back off, look at the geographical pattern, and test whether this could really have been an old valley of the Tiber.

The canyon of the modern Treia River runs from south to north, from Mazzano to Civita Castellana, and in the walls of the canyon we can begin to see what happened when the Tiberina ignimbrite flowed into the old valley.

The river gravels are higher in elevation in the north, and they gradually descend toward the south, which has the same slope as the modern Tiber, far away to the east. From Mazzano to Narce a thick body of Tiberina ignimbrite rests on the gravel. North of Calcata the ignimbrite is thinner, and is overlapped by lake sediments. Still farther north the ignimbrite disappears and the lake beds rest directly on the gravel. It is quite clear from these relations what happened. The ignimbrite was erupted from the earliest of the Sabatini volcanoes, south of Mazzano, and it flowed northward up the old river valley, getting progressively thinner. This ignimbrite dammed the Paleotiber Valley, and a natural reservoir was impounded behind the dam.

So far so good, but still a question remains: Why didn't the lake keep filling until it spilled over the ignimbrite dam and eroded it away? That is what normally happens in cases like this. In the Grand Canyon, for example, the Colorado River was dammed many times by lava flows that were soon eroded away when the resulting lakes spilled over the dams.[15] The special circumstances of the Paleotiber become clear when we plot the situation on a map.

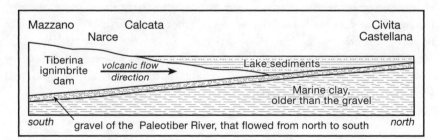

In the walls of the modern Treia Canyon the rocks show us how the volcanic flow of the Tiberina ignimbrite dammed the ancient valley of the Paleotiber, impounding a lake where sediments were deposited.

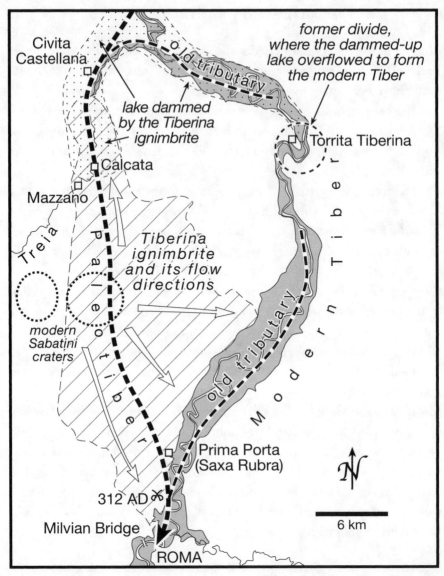

This map shows how the Paleotiber River was diverted, about 550,000 years ago. The Tiberina ignimbrite, erupted from the first volcano of the Sabatini district, dammed the Paleotiber so that a lake backed up an old tributary valley. The lake eventually spilled over a drainage divide at Torrita Tiberina into a second tributary of the Paleotiber, which thus became the route of the modern Tiber.

On the map we can see how an ancient valley ran south from Civita Castellana through Calcata to Prima Porta. It is clear that this valley was the path of the Paleotiber, following a shorter route than its modern detour around to the east. If we examine the modern Tiber, we see that its valley is abnormally narrow near Torrita Tiberina, where there are some peculiar sharp bends in the valley.

My interpretation is that the Paleotiber had a couple of old tributaries whose headwaters were separated by a drainage divide at Torrita Tiberina. When the ignimbrite dammed the Paleotiber, the lake behind the dam gradually filled back to Civita Castellana, then up the northern tributary valley. Evidently the divide at Torrita was lower in elevation than the ignimbrite dam, and the lake eventually spilled over the divide into the valley of the southern tributary. The overflow cut the narrow, curving channel at Torrita, and the southern tributary became the main valley of the modern Tiber.

Thus the Paleotiber was at one time dammed and dead, with little or no water where its channel passed through the future site of Rome. We can even say just when that was, for Dan, Fabrizio, and Paul have dated the Tiberina ignimbrite as about 550,000 years old. It probably took only a few decades for the Paleotiber to fill the natural reservoir to the point where it spilled over the divide at Torrita Tiberina. At that moment the Tiber was reborn with a new route, renewing the flow of river water through what is now the city of Rome.

A surprising event in landscape history, like this damming and diversion of the Tiber, has its own fascination. But does it have any relevance to human history?

Consider this: The landscape around Prima Porta was sculpted by the Tiber, coming in from the northeast. Had the drainage divide at Torrita Tiberina been slightly higher, the Tiber would have overflowed the ignimbrite dam, resumed its original route, straight south through the Sabatini volcanoes, and the landscape of Prima Porta would have been different. This is significant, because one of the battles that truly changed history took place just south of Prima Porta, which in Roman times was called Saxa Rubra, or "Red Rocks," after the red ignimbrite of the *Tufo rosso a scorie nere*.

In A.D. 312, Saxa Rubra was the site of Constantine's famous nighttime vision of the cross, with the words "In this sign you shall conquer." The vision came on the eve of the Battle of the Milvian Bridge, in which Constantine was victorious over his rival, Maxentius. The vision and the victory led to the establishment of Christianity as the official religion of the Late Roman Empire.

Had the Tiber not been diverted, and had the landscape between Prima Porta and Rome thus been different, and had the battle still taken place there, would Constantine still have been the victor? The outcomes of battles are notoriously dependent on the most subtle chance events, and a different landscape might very well have led to a different result. Had the Tiber not been diverted, the Roman Empire might not have become Christian. It is difficult to imagine how different the world would have been, from then to now.[16]

A Farewell to Rome and Its Volcanoes

It is time now to leave Rome. This has been a perfect place to learn the fundamentals of reading Earth history written in rocks, for the Roman volcanics and the Tiber sediments are very young, and their relation to the changing landscape is easily understood. In Rome we encountered a lost world, but just barely lost, for surely human eyes once looked upon a pristine landscape in this place.

All the geology we have seen so far is younger than the symbolic "million years"—the forbidding gateway to Earth's deep past. Now it is time to go back farther into time.

Rome and the Treia Valley have also given us the opportunity to see some of the ways geologists go about reading the rock record, by studying outcrops and cores, drawing stratigraphic columns, constructing geological cross sections, and plotting the information from rocks on maps. Now that we know some of the tools and are perhaps comfortable with ages in tens or hundreds of thousands of years, it is time to cross that symbolic million-year barrier.

We can see just a hint of a more ancient world along the banks of the Treia River at Narce, where the cross sections show something underneath the gravels of the Paleotiber. That something is a clay containing fossil shells of animals that only live in the seas—never in fresh water. This is our first hint of a whole different chapter in the history of Italy, a time before the Paleotiber, when much of the present peninsula was below sea level. We get only a tiny glimpse of this ancient world at Narce. To see it better we need to get away from the Roman volcanics, to a place where the older sediments are more fully exposed. That place is Tuscany, and there we will begin to appreciate the geological timescale—the framework for understanding Earth history.

PART III

——✦——

Siena and Gubbio

In which we travel to Siena and Gubbio to explore the second great question of the book—How can we determine the ages of rocks and the ages of the events in Earth history that the rocks record? We shall see that generations of geologists gradually discovered how to find the ages of rocks in a variety of ways, so that now we can build a historical chronology of the Earth's past, placing in sequence and in causal relationship all the long-lasting trends and chance events that have built the unique Earth we know today.

5

SIENA AND THE DISCOVERY OF EARTH HISTORY

ROME AND ITS SURROUNDINGS have given us an opportunity to see how geologists explore and reconstruct the ancient, lost worlds of Earth's past. Rome and its sleeping volcanoes were particularly suited to answering this first of the three grand questions of the book, because active volcanoes are familiar and visible features of today's landscape, known to most people through pictures, if not through actual experience. Rome was also a good place to start because its volcanic past is geologically so young—less than a million years old.

When we come to explore the Apennine Mountains in part 4, we will encounter lost worlds that stretch back *hundreds* of millions of years into the past, and landscapes that are strange and unfamiliar because modern examples lie hidden deep in the ocean. This kind of journey into deep time requires some preparation. We need to learn how geologists determine the ages of the rocks that remember those ancient worlds. Without the ability to date the rocks, we would simply be lost in a wilderness of time, surrounded by marvels that made no sense at all, because we could not see how any one past event related to any other.

And so in part 3 we address the second of the grand questions of the book—How can we determine the ages of rocks, and of the events in Earth history that the rocks record? This question may at first seem rather technical, but it is the key to appreciating Earth history. Think about how hopeless your understanding of human history would be if you knew about various historical episodes—like the Fascist dictatorship of Mussolini, the building of the Aurelian Wall of Rome, and the life of Saint Francis—but had no idea when, or even in what order, they happened.

The same would be true if you could not place in their proper order the sedimentation of the blue clays of Tuscany, the deposition of the *Scaglia* limestone of the Assisi quarries, and the volcanic eruptions of Rome. Dating is crucial! To see how Earth history is dated we will first visit Siena and then go on to Gubbio. In these two remarkable little medieval cities we will be able to trace the steps by which geologists gradually developed a timescale for Earth history and learned how to tie past events into that chronological framework.

The Pilgrimage Road to Siena

The Treia Valley afforded no more than a peek at the sediments that lie beneath the Roman volcanics, but on our journey to Siena we will enter the domain of those older deposits, where there is no volcanic cover to hide them. So let us return from Mazzano to the Via Cassia, and instead of heading back south to Rome, let's now follow the Cassia northward, through the volcanic countryside. Our route takes us close to the volcanic crater lake of Vico, through the little city of Viterbo (where the popes sometimes retreated when things got too nasty in Rome), and then along the shore of another volcanic lake, Bolsena, which marks the northernmost of the four big volcanic centers flanking Rome.

The Via Cassia passes the northern limit of the volcanic rocks at Acquapendente, our gateway to the older geology of Italy. Now we find ourselves in a completely different landscape—in a realm of yellow sands and blue clays, deposited near the ancient coastline as Italy rose out of the sea, before the volcanoes began to erupt.

We have now passed the symbolic gateway into deep time that lies at

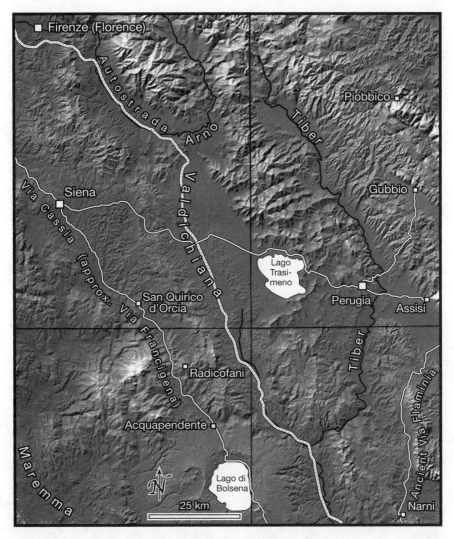

Tuscany and Umbria, north of the Roman volcanic region, showing the route our journey will follow, along the Via Cassia to Siena, and then across the ranges of hills to Gubbio.

the date of one million years ago. Passing that barrier marks the key to thinking like a geologist, to whom a million years is the basic unit of time and one million years *ago* represents the very recent past.

The yellow sands and blue clays after Acquapendente date from the time geologists call the Pliocene—just a little older than the symbolic one-million-year barrier—so they will let us ease into the abyss of deep time. These sediments are so young that they have not yet solidified into hard rocks. Being weak and easily broken down to form soil, they account for the gentle hills and soft agricultural landscapes that make this part of Italy so beautiful.

Our route takes us past evocative medieval towns like Radicofani and San Quirico d'Orcia, as we pass from Latium, the land of the Latins, into Tuscany, the land of the Etruscans. Ahead of us lies one of the most richly medieval parts of Italy, so perhaps it would be good to stop thinking of this route as the Via Cassia, its name in ancient Roman times and again today, and think of it instead as the Via Francigena (pronounced fran-CHEE-gena), as they called it in the Middle Ages. At that time it served as the main route from England, France, and Germany to Rome, the center of Western Christianity, and the name means "the road (to Rome) coming from France."[1]

Traveling north on the Via Cassia, we eventually reach Siena, a little architectural masterpiece of city, where a visitor finds the spirit of the

The Cathedral of Siena, with its dome and tower, was intended as the transept of a much larger church, whose unfinished end wall, with two tall arches, stands alone at the left.

The Piazza del Campo of Siena, with its graceful tower rising above the Palazzo Pubblico, from which Siena was governed during the days of her greatest glory. Twice each summer the piazza is the site of the medieval horse race called the Palio.

Middle Ages still in full flower. Siena crowns a hill of Pliocene sediments, culminating in a magnificent cathedral. Despite the great size of this church, the Sienese intended it as only the transept of a truly enormous cathedral. The nave of the larger composition had been partially completed when it had to be abandoned because of the catastrophe of the Black Plague, which reached Siena in 1348.

From the top of the unfinished end wall of the still larger church, we can look down into the nearby Piazza del Campo, with its elegant Palazzo Pubblico, culminating in the graceful tower called the Torre del Mangia. Twice each summer the Sienese bring earth into the Piazza del Campo to make its perimeter into a racetrack, and on that track they hold the wild horserace called the Palio, honoring the Madonna on two of her important feast days.[2]

Wandering through the elegant streets of Siena, you cannot help wondering how this little city, so far off today's beaten track, became wealthy enough to adorn itself with such lovely buildings on such a magnificent scale. Much of the wealth came from travelers following the Via Francigena to Rome.[3] Of more interest to a geologist, an important mining district lay within the territory of Siena. The *Colline Metallifere*—the "metal-rich hills"—produced iron, lead, zinc, silver, copper, antimony, and mercury, from at least as far back as the Middle Ages until the twentieth century. Ore deposits are often related to volcanic districts, and Italian geologists have shown that the metal ores that helped make Siena rich came from the igneous activity that produced some small volcanic areas and granite bodies in southern Tuscany, just before the much larger eruptions of the Roman volcanoes.[4]

Here in Siena and in nearby parts of Tuscany, we will meet two early geologists. The first, Nicolaus Steno, who lived in the seventeenth century, has long been familiar to geologists, who consider him the founder of our science. The other, Ambrogio Soldani, born a century after Steno, is almost unknown among geologists, but since I learned about his discoveries, he has joined the pantheon of my personal scientific heroes. Nicolaus Steno and Ambrogio Soldani will lead us into the question of how geologists learned to determine the ages of rocks.

Nicolaus Steno in Tuscany, and the Discovery of Earth History

In coming to appreciate deep time, measured in millions of years, it will help if we understand the very difficult changes in thinking that led the first geologists to recognize the enormous age of the Earth. It is especially useful today, when even in a scientifically advanced country like the United States many people believe that Earth history has lasted no more than a few thousand years, based on biblical literalism. So we need to understand how the evidence in the rocks forced the early geologists, against their will, to reject the short biblical timescale that dominated thinking at the time.

During the Middle Ages religious faith underlay virtually all thinking in Europe. The Renaissance saw the emergence of humanistic but not very scientific thought. During the Reformation questions of faith convulsed Europe until the end of the Thirty Years' War in 1648. That brings us to the late seventeenth century, the time of the first great flowering of science, symbolized by Isaac Newton's *Principia Mathematica*—the Mathematical Principles of Natural Philosophy— published in 1687. Newton's mathematical laws made complete and predictable sense of the observed motions of sun, moon, and planets, gaining enormous prestige for his new physics. However, these mathematical laws in no way interfered with Christian beliefs, strongly held by Newton himself.

Nicolaus Steno (1638–86), who discovered in Tuscany that Earth's history is written in rocks.

It took a much longer and more complicated effort to uncover the Earth's past than to work out the laws of physics. Unlike the motions of planets in the night sky, the history of the Earth could not simply be observed. In Newton's time the only known information on the Earth's past came from the Book of Genesis, which recounted genealogies suggesting that the creation of the Earth had taken place just a few thousand years before.[5]

Only a remarkable scientific genius could have realized that Earth has a history recorded in rocks, and could have figured out how to read that history. That genius was Niels Stensen, born in Denmark in 1638, although he followed a custom of scholars of the time and Latinized his name to Nicolaus Steno. Excelling in anatomical dissection, Steno developed his scientific abilities in Amsterdam and Paris before arriving in Florence, attracted by the vibrant scientific life centered around the court of Grand Duke Ferdinando II de' Medici and his brother, Prince Leopoldo.[6] His background in anatomical research gave Steno the tools to become the first anatomist of the Earth, but it was his particular genius to realize that the earthly anatomy he was uncovering was a record of history.

The Fossils of Tuscany

As we followed the Via Cassia northward toward Siena, through hills of Pliocene blue clay and yellow sand, we passed many places where we could have easily have found fossil shells. Each spring after the winter rains, a whole new crop of fossils appears, washed out of the soft clay and sand that enclose them. It would almost seem as if the fossils were growing inside the sediments, and for a long time most people who thought about these matters mistakenly believed just that. Let us see why:

The sequence of creation in Genesis raised serious problems for anyone from the Renaissance to the seventeenth century who wondered about fossil shells.[7] The fossils in Tuscany looked very much like the shells of animals that live today in the sea off the Tuscan coast, but how could the sea ever have covered today's dry land, on an Earth with only a few thousand years of history? Some concluded that the fossils represented animals that had lived when Noah's flood covered the mountaintops, but this view raised serious problems. For example the Genesis account places the creation of the Earth long before Noah's flood, so how could fossil shells from the time of the flood get *inside* rocks that formed *earlier*, when the Earth was created?

In the sixteenth and seventeenth centuries contradictions like this seemed to rule out the notion that fossils had ever been living animals. Strange as it seems to us today, most thinkers of the time preferred to conclude that the fossils had grown inside the rocks through the mysterious action of unknown forces, for unknown reasons. They were thought of as "sports of nature," and maybe that was acceptable because so little was then understood about what nature can and cannot do.

In Florence in 1666 Steno had the opportunity to dissect the head of a giant shark caught off the Tuscan coast, and he could clearly see detailed similarities between the teeth in the shark's jaw and fossils called "tongue stones," found in rocks around the Mediterranean. This realization led Steno to explore the hills of Tuscany with the aim of really understanding the origin of fossils.

Tuscany was the perfect place for his quest, and the patronage of the Medici brothers opened the stone quarries and the mines of the *Colline Metallifere* to his investigations.[8] Although Siena had by that time been incorporated into the domains of the Medici, it appears that Steno did not visit Siena.[9] Nevertheless the Pliocene clays and sands of the Tuscan hills near Siena allowed him to make observations that not only proved that fossils are the remains of animals and plants but provided the methods for reading Earth history written in rocks. In this sense Nicolaus Steno invented the science of geology in Tuscany in the 1660s.

I doubt if anyone had ever before made such thoughtful observations of rocks, although that is exactly what field geologists do today.[10] Steno began by demonstrating that fossils do *not* grow inside the rocks. The evidence was compelling. He never saw the cracks in the surrounding rock that a growing fossil would surely make. He found many fossils in the process of falling apart—quite the opposite of growing. He noted that fossils never had twisted, distorted shapes, like the roots of trees that *do* grow inside soil. He found many broken fossil shells, and sometimes found two nearby parts that fitted together perfectly. And he found clusters of fossils with just the same arrangement as clusters of shells in today's seas.[11] There could be no doubt that the fossils represented the shells of once-living animals, entombed in soft sediments that accumulated in the places where the animals had lived.

Layered Solids and the Order of Historical Events

Understanding the nature of fossils was important, but it led Steno to an even more important realization. He came to understand that solid bodies of any kind do not suddenly appear, fully formed. Instead they are constructed, bit by bit, and often they are constructed in layers. We commonly see layers of stone blocks in buildings, and in the photo of the Cathedral of Siena we see the layers emphasized by bands of different-colored stone, as if to remind us that layered history was discovered in the nearby hills. Layered history is also familiar from tree rings, coats of paint on old buildings, and layers in fossil shells.

Layers thus represent time and history, and, most important for the geologist, the layers of bedded sedimentary rocks represent Earth history. This was perhaps Steno's greatest discovery, and geologists call it the law of superposition. Sometimes it is stated in a simplified way, something like

this: "If a rock is on top, it is younger (unless there has been a complication, like the filling of a cavity, or overturning)."

In those terms the law of superposition seems almost laughably obvious, but it is worth our attention for three reasons. First, apparently no
one had ever pointed it out before Steno did. Second, it is the basis on
which the history of our Earth has been reconstructed. And third, Steno's
formulation of the law was actually much more subtle than the caricature
version that most geology students learn.[12] Given the sophistication of his
understanding, it would be fair, I think, to say that Nicolaus Steno laid the
foundation for today's critical field of condensed-matter physics, as well as
of geology, although it would probably be rare to find a physicist who has
ever heard of him.

With the law of superposition Steno gave us our first handle on the
chronology of events in Earth history. It was still not possible in any mean-
ingful way to determine the ages of events, but superposition allowed Steno
and those who followed him to determine the *order* in which events had
happened. To apply this concept to our previous case, it was now possible
to show that the deposition of the yellow sands and blue clays of Tuscany
came before the volcanic eruptions of Rome, because the volcanic rocks
rest on top of the sands and clays at Acquapendente.

With the first chronological tool of geologists in place, let us see how
Steno immediately used it to do the first deciphering of some Earth history.

Working Out the Geological History of Tuscany

Nicolaus Steno, with his anatomist's eye for geometry, drew some remarkable conclusions about how the landscape of Tuscany had come to be.
First of all, once he had demonstrated that fossils really do represent the
shells of marine animals, they gave him the evidence that at one time the
sea had indeed covered parts of what is now dry land. Second, Steno realized that layers now tilted had once lain flat and horizontal. And third, he
understood that younger sediments had accumulated in valleys bounded
by tilted, older beds.[13]

This third conclusion cuts to the heart of the mystery of the flat Apennine valleys that puzzled me during our winter trip to Assisi. I was then
thinking that all mountain valleys should be steep-sided erosional gorges

How to get tilted beds and Tuscan valleys filled with sediment

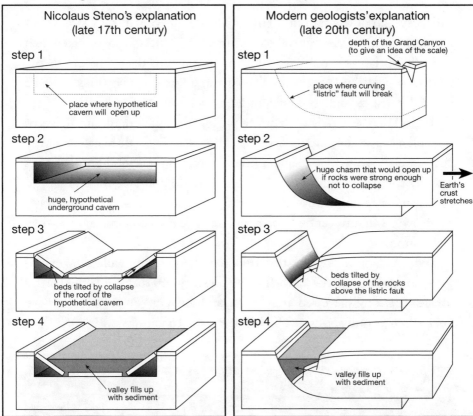

These diagrams show how remarkably Steno's seventeenth-century explanation for tilted beds and sediment-filled valleys (left) anticipated the explanation preferred by geologists today (right). In the figure on the right, steps 2, 3, and 4 all happen together, but it is helpful to separate them diagrammatically to show the close parallel to Steno's thinking.

where mountain streams cut into bedrock. But Steno understood, three hundred years earlier, that if the Earth's crust collapses for some reason, the resulting depressions will fill up with sediments, producing wide, flat-bottomed valleys.

Steno used his new discoveries to draw the very first geological cross sections ever made. It is worth looking very carefully at them, for they allow us to see through the eyes of a remarkable genius, and they anticipate in a quite astonishing way the view of modern geologists.

Nicolaus Steno's diagrams at first look a bit strange to our eye, for he drew them in a very diagrammatic, unnatural way, with ruler-straight planks instead of the naturally irregular contours of real rock strata. But here again Steno was far ahead of his time, for scientists today make much use of "models," stripping away all the extraneous detail, to get to the heart of a question.[14] Steno's very diagrammatic cross sections must be one of the first examples of a scientific model.

To explain the tilted beds Steno envisioned giant underground caverns that collapsed so that the roofs tilted as they fell. Early thinkers about the Earth often invoked great underground caverns, and modern geologists consider them a quaint fantasy, because rocks do not have the strength to support huge caverns.[15] So it is ironical that today's explanation invokes great caverns that never quite open up, because they collapse even as they are forming!

Today geologists explain the Tuscan and Apennine valleys by extension—the pulling apart—of the Earth's crust. When rocks near the Earth's surface are strongly stressed, they break, and the rocks on either side of the break slip past one another. The break is called a fault.

A vertical fault exposed in a quarry at Furlo offsets the thick white bed that dips toward the left.

Modern geologists have discovered that when Earth's crust is pulled apart, the stretching happens by slipping along "listric" (spoon-shaped) curving faults that flatten with depth. From step 2 on the right side of the diagram, it is clear that extension on a listric fault *would* open up a giant cavern, if the rocks were strong enough. But rocks are weak, so the ones above the listric fault collapse continually, and the cavern never opens up. Step 3 on the right shows how the collapse tilts the beds.

Listric faults definitely produce tilted beds and sediment-filled valleys. Our modern understanding does not differ fundamentally from Steno's if we allow him two adjustments—first, his caverns resulted from extension of the crust, and second, they collapsed progressively as they formed, so that no huge cavity in the ground ever existed.

In addition to explaining the tilted beds, Steno correctly understood that younger sediments partly filled the collapsed valleys, and that the beds of step 4 accumulated in the low areas of a mountainous landscape, when the sea flooded the dry land. He wrote that the diagram "shows new strata, made by the seas, in the valleys."[16]

One of Steno's geological successors at Florence, Francesco Rodolico, pointed out in 1971 that Steno was accurately recognizing what we now call the transgression, or flooding, of the Pliocene Sea, leading to deposition of the blue clays and yellow sands of the Via Cassia.[17] Steno got that part exactly right, but neither he nor Professor Rodolico could possibly have imagined the amazing story behind the Pliocene flooding. It was not the flood that was strange—the *really* strange thing was the dry land that preceded it. That dry land existed because just before, during a time interval called the Messinian, the Mediterranean Sea had largely dried up. This astonishing story will emerge in chapter 13. The Pliocene flood occurred not because the sea overflowed but because it refilled its normal basin.[18]

Let us make an evaluation. How well did Steno do? He accurately recognized two cycles of collapse and sedimentation, although I have really only talked about the first one. Steno had only the Bible as a guide to dating his two cycles of sedimentation, so he attributed the first one to the universal ocean of Genesis, before the appearance of dry land, and the second one (step 4 in our diagram) to Noah's Flood. This dating is wrong, of course, for the discovery of deep time would not come until later, but

Nicolaus Steno really did figure out how to read history in rocks. He is deservedly the founding hero of the science of geology.

Unfortunately we cannot know in detail the field observations and reasoning that lay behind Nicolaus Steno's remarkable cross sections of Tuscany. He published them in 1669 in a brief *Prodromus*—a preview of a longer "dissertation" that he planned to write.[19] However, at that same time Steno converted from his native Lutheranism to the Catholicism of Florence. The Church made him a bishop, and he spent the rest of his life as a missionary to the Protestants of northern Europe. In 1988, after an investigation of his life and works, the Roman Church held a ceremony of beatification for Nicolaus Steno.

When Steno turned to the religious life, he stopped his study of nature and never wrote the promised dissertation. We will never know how much he really understood about Tuscan geology, but the *Prodromus* alone tells us enough about his thinking to establish his fundamental role in the history of science. Geologists have not forgotten. In August 2004, at the International Geological Congress in Florence—the first one in Italy in more than a century—my friend Gian Battista Vai led a pilgrimage to Steno's tomb, in the Basilica di San Lorenzo in Florence, to pay homage to the founder of stratigraphy and the discoverer of the historical record of the Earth—a scientific genius who has been elevated the first step toward becoming a saint.

Ambrogio Soldani, Fossil Animals, and the Age of Rocks

On our most recent trip to Siena, on a crisp March morning, Enrico Tav-arnelli told Milly and me that he had something special to show us. Enrico is a very enthusiastic young geology professor who had been a postdoc-toral fellow in Berkeley and had loved our university. Now he was eager to show us some little-known treasures of his native Siena.[20]

Enrico led us to a well-hidden old monastery, dating back to the twelfth-century glory days of the city and blending into the surrounding medieval buildings. This had once been a house of monks of the Camal-dolese order, whose abbot in the second half of the eighteenth century was a remarkable man named Ambrogio Soldani. In 1816 the monastery, no longer in use by the monks, became the seat of the Accademia dei

Fisiocritici—the "Academy of the Critical Thinkers about Nature." This is the natural history society of Siena, founded all the way back in 1691, of which Ambrogio Soldani had been the secretary.[21]

Ushering us through the doors of the old monastery, Enrico introduced Milly and me to the geologists of the Accademia, who were eager to show us their collections of fossils and minerals, dating back to the days when no one understood very much about the Earth, and geology was just beginning to be a science. Wandering through the Accademia, which today is a modern natural history museum as well as a historical repository, the geologists showed us cases full of fossils from the Pliocene blue clays and yellow sands around Siena.[22] These fossils were just like the ones that led Steno to his great intellectual breakthrough, and the collections were started just decades after Steno's pioneering work.[23]

In a way that I could never have imagined, this monastery was about to disclose a direct link between the pioneering geologic research of Nicolaus Steno in the hills around Siena and the geologic research we did at Gubbio more than three hundred years later.

Steno used fossils to demonstrate that marine waters had once covered places now above sea level. If fossils told no more of a story than that, they would merit only a footnote in the history of science.

But we now know that there was a much greater discovery about fossils waiting to be made—the law of faunal succession—that the shapes and forms of fossils are different in rocks of different ages, with the different shapes and forms appearing in a particular order going from older to younger. The law of faunal succession was to be the key to dating Earth history with fossils—the next of the major methods by which geologists unravel Earth history. Fossils make it possible to go beyond simply arranging Earth-history events in their order of occurrence on the basis of superposition—they make it possible to tie the events into an actual chronological framework.

Would Steno have discovered faunal succession himself if he had continued to investigate Earth history in Tuscany? He clearly understood that younger rocks rest on older ones, so he had that part ready to go, and he was making careful observations of fossils with his anatomist's eye. But showing that older fossils have different shapes from younger ones would have been hard for Steno, because the abundant fossils in the Pliocene blue clays and yellow sands of Tuscany are so recent that they have almost exactly the same shapes as the shells of animals living in the sea today.

In Tuscany there are plenty of sedimentary rocks old enough to contain fossils that would look very different, but most of those rocks happen to contain almost no large fossils at all.[24] I suspect that Steno looked at them, saw no fossils, and therefore attributed them to the first day of Creation, before animals were created.[25] Had Steno continued to focus his attention on geology instead of on religion, he might have come across one of the few places in Tuscany where old fossils are abundant. Or perhaps he might have visited the high Apennines to the east of Siena, where there are many such places. But Steno did not continue to study the Earth, so we will never know if he would have discovered fossil succession.

The most abundant and useful of the old, large fossils in this part of Italy are the ammonites—disk-like spiral shells, from the size of a coin to that of a fist—that are related to today's chambered nautilus but are now entirely extinct. There are a couple of small collections of ammonites among the plethora of much younger Pliocene fossils in the Accademia dei Fisiocritici, reflecting their rarity in Tuscany, but to the east, in the Mountains of Saint Francis, ammonites are abundant and spectacular. Ammonites are just about the most perfect fossils for dating rocks. They floated in the surface waters of the oceans, so any evolutionary changes were quickly spread all over the world, and they evolved very rapidly, so their shapes and structural details are diagnostic of specific, brief intervals of time. In rocks that contain ammonites the problem of dating is solved. Unfortunately there are many, many rocks in which these precious chronometers are absent. Let us trace some threads of history and see how geologists found other ways to date rocks.

Discovering a New Hero

The scientists who came after Steno and debated the origin and significance of fossils and their changes through time mostly thought of themselves as "naturalists"—as students of "natural history." Those names appeal to me, but they have fallen into disfavor and suggest an old-fashioned approach to science. None of the naturalists of that intellectual effort, from the mid-seventeenth to the mid-nineteenth century, are widely rec-

Ferruccio Farsi, Giovanni Guasparri, and Roberto Fondi—geologists of the Accademia dei Fisiocritici of Siena, standing behind the polished case containing the collection of microfossils of Ambrogio Soldani, the founder of micropaleontology.

ognized today, until we come to Darwin. Few of today's geologists, their intellectual heirs, have heard of more than a few of them. Of the many who were active in Tuscany,[26] let me pick just one representative whose discoveries will lead directly into the Earth history we will explore in the next chapter. That representative will be Ambrogio Soldani, the Camaldolese abbot of Siena and secretary of the Accademia dei Fisiocritici.

In one room of the Accademia the geologists showed us a large and elegant case of polished wood with an inscription declaring that it contains the microscopic fossils collected by Abbot Soldani. As I admired this case I realized that it marked the beginning of a long line of scientific investigation that led to my own work in the Apennines.

People wandering around in the hills have noticed fossils—big, robust ones—probably back to the beginning of human curiosity. Only more recently have people noticed and studied tiny fossils, at or below the resolution of the human eye. Yet these microfossils have played a major role in bringing to light the history of the Earth. Microfossils called Foraminifera ("forams" for short), barely visible in the pink-and-white limestone of the Basilica of Saint Francis at Assisi, eventually led us to an understanding of impacts and mass extinctions, and to the discovery of a giant impact crater buried beneath the surface in Mexico.

So I realized with great pleasure that Ambrogio Soldani had invented micropaleontology—the investigation of microfossils—right here at the Accademia dei Fisiocritici in Siena.[27] Abbot Soldani spent more than forty years separating microfossils from the Pliocene sands and clays around Siena and studying them under the microscope.[28]

Soldani thus pioneered an absolutely critical technique in geology. It may be hard to see the tiny forams without a microscope, but they offer great advantages for determining the ages of rocks. Being so small, they can also occur in great abundance, and you can sometimes find thousands in a single handful of sediment. This gives geologists a better chance of finding the few critical species for dating the sediment. Oil geologists can

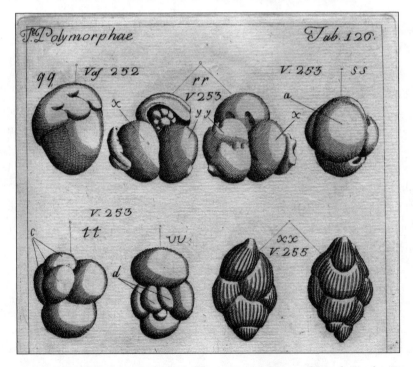

Microfossils called "forams," from the Pliocene around Siena, drawn by Ambrogio Soldani as he saw them under his microscope.

recover large numbers from well cores that would be very unlikely to bring up a single, undamaged large fossil, so forams have been widely used in the petroleum industry and have been critical in finding large amounts of oil.

Some forams live on the sea floor, but others float in the surface waters until they die, and then their shells sink to the bottom and end up in the sediment. As with ammonites this helps geologists, because the first appearance of the new species in sediments from different places should be close to an ideal time marker.

The last appearances of species, marking their extinction, also provide good time markers. In the Apennine limestones, the first and last appearances of forams are scattered apparently randomly through time, with one exception—a level where almost all the species of forams disappear at exactly the same time. This "mass extinction" marks the historical milestone that geologists call the Cretaceous-Tertiary boundary.

Not only forams but many other groups of plants and animals died out at that same time. At Gubbio—the next place we shall visit on our field trip—the Cretaceous-Tertiary mass extinction of forams recorded in the rocks drew the attention of our Berkeley research group in the 1970s. Our studies led to the 1980 hypothesis that a giant comet or asteroid had collided with the Earth and caused the mass extinction. Eleven years later the recognition of a giant crater hidden below the surface in Mexico and dating from exactly the Cretaceous-Tertiary boundary confirmed the impact explanation for this mass extinction.

Seeing the polished case with Ambrogio Soldani's foram collection in Siena thus made a dramatic connection for me. It placed our work at Gubbio in a historical context, and it let me recognize that our work followed in the footsteps of this man whose name I had never heard. It started me thinking of Soldani as a scientific predecessor whose memory I could revere, and whose accomplishments I could tell my friends about.

Then, in the very next hall of the Accademia dei Fisiocritici, the connection suddenly became even more remarkable. Ferruccio Farsi, Giovanni Guasparri, and Roberto Fondi—geologists from the museum—pointed out a rather nondescript-looking rock with a label that said "Meteorite di Siena," and they told me its story.

Just as Ambrogio Soldani was concluding his studies of microfossils, they said, a dramatic event diverted him into a new field of science. At

Ambrogio Soldani (1736–1808), abbot of the Camaldolese monks at Siena, pioneer of micropaleontology, and early advocate for the extraterrestrial origin of meteorites. This anonymous portrait from the Accademia dei Fisiocritici was painted in the early nineteenth century, near the end of the abbot's life, and shows him with his geologist's hand lens and some of his drawings of microfossils.

about sunset on the evening of June 16, 1794, the sky south of Siena lit up, to the accompaniment of "extraordinary explosions," and flaming stones fell to Earth. Fascinated, Soldani assembled a collection of the fallen stones at the Accademia and set about studying them.

After carefully evaluating the event, Soldani argued forcefully that the stones were of extraterrestrial origin,[29] a view that brought down upon him a barrage of ridicule from other prominent scientists of the time. Analyzing the fallen stones presented great difficulties, because modern chemistry was just coming into existence. Soldani gave samples to so many chemists that only one small piece of the Siena meteorite still remains in Siena. When the chemists had finished, their results supported

his extraterrestrial interpretation, which has held up and been strength-
ened ever since. Wouldn't Soldani have loved to see the impact of Comet
Shoemaker-Levy-9 on Jupiter in 1994, exactly two hundred years later?

It is sad that almost nobody outside Siena remembers Soldani,
for scientists today should honor the name of this early geologist. In
Ambrogio Soldani I have found a new scientific hero, for he pioneered
the study of both forams and meteorites—studies that came together at
Gubbio, almost two centuries after his death, in the impact theory for
the Cretaceous-Tertiary mass extinction.

In Siena we have traced the beginning stages in how geologists learned to
date the events in Earth history, using the stacking order of sedimentary
layers and the evolutionary changes in the fossils they contain. The time
has come to leave Siena. Let us pack up and head east, continuing on our
field trip, through the valleys of Nicolaus Steno and into the Mountains
of Saint Francis. Our journey will bring us to Gubbio, another wonderful
little medieval city. There we will see how both branches of the pioneering work of Ambrogio Soldani came to fruition and how geologists learned
not only how to decipher the history written in rocks but how to assemble
all the events of Earth history into a dated chronological sequence.

6

GUBBIO AND THE CHRONOLOGY OF THE PAST

THE WIND WHIPPED through lifeless streets the first time Milly and I ever saw the medieval city of Gubbio. We had left Assisi that morning, on our Christmas trip of 1970, and as we crossed the barren Apennine foothills, the weather got worse and worse. Shivering beneath a harsh sky, Gubbio brought a whole new meaning to the word "bleak." Gray stone buildings stood huddled together, with claustrophobic little arched entrances called the "doors of the dead." Ancient cobblestone streets were piled with dirty snow and slick with ice. In Assisi we had learned that Saint Francis once came here to persuade a ferocious wolf to stop eating the town's children, and as the wind howled through Gubbio, we could almost believe the wolf still stalked its streets.

Never had the Middle Ages seemed so alien and menacing, and it was a relief to depart through one of the gates in the city walls. We started up into the mountains, through a canyon called the Bottaccione Gorge, and into a gathering storm. I doubt that either of us imagined we would ever return—let alone that Gubbio, and this canyon, would become the center of our lives.

I do not remember noticing the rock outcrops along the Bottaccione road at the time, and they were probably covered with snow. I did not know that this canyon was already of considerable importance in geology, or that it was destined to become far more so. Unaware, we were driving past rock formations whose names were soon to become second nature— *Majolica, Scaglia, Bisciaro.* Pronounced in the Italian way, these names even have a romantic sound—ma-YO-lee-kah, SKAL-yah, be-SHAH-ro. They are about to become major characters in our story.

The Bottaccione Gorge will be a good place to carry on the themes of the Siena chapter and to see how geologists gradually developed a whole suite of methods for dating sedimentary rocks. Using those tools, geologists now can reconstruct Earth history in great detail, as we will do for the Apennine Mountains in the final part of the book.

Guido Bonarelli's Adventures and Frustrations

Most of the rocks we were passing that winter day in the Bottaccione Gorge are limestones—white, gray, pink, or red in color—but at one place there is a unique bed of jet black shale and chert just a meter thick.[1] The Italian geologists call this distinctive bed the Bonarelli level, in memory of a remarkable man. I had not yet heard of Guido Bonarelli, but now that I know about him, I really wish we could have met. He surely could have told some wonderful stories.

In 1871 Guido Bonarelli was born to a titled family in nearby Ancona, but his family moved to Gubbio when he was very small. Gubbio was always the home to which he would return between his remarkable journeys. Somehow he was attracted to geology at a time when this was a very rare passion. I can only think that like so many of us a century later, he fell under the spell of the rocks of Gubbio, for his first publication, when he was only twenty, was titled *The Territory of Gubbio: Geological News.*[2] In this paper he first described the bed of black shale and chert that now bears his name.[3]

Young Bonarelli was soon on track to becoming a successful academic geologist, receiving a professorship in 1897 at Perugia, just twenty miles from his beloved Gubbio.[4] It would have seemed the perfect place to put down roots, but Bonarelli's vision and horizons extended far beyond this part of Umbria. In 1901, at age thirty, he resigned from the university and left Italy for the remote and wild island of Borneo to apply his geological skills to the search for oil. Borneo was then a colony of the Netherlands, and Bonarelli went to work for the Royal Dutch Petroleum Company. Geology has always been a wonderful excuse for adventure, and I imagine travel was part of the attraction. Like many another geologist, he was fascinated by the people he met on his travels, and several of his 158 publications dealt with anthropology.

But Guido Bonarelli must also have had a remarkable foresight about what was to come as the twentieth century unfolded, as national economies came to be powered by oil and natural gas, with wars and diplomacy driven by the need for these critical resources. The geological search for petroleum was to be the focus of much of the rest of his life. Bonarelli pursued this quest in many parts of the world, notably in South America, where he pioneered the geological exploration of Argentina, including Patagonia and Tierra del Fuego.

Bonarelli represents the practical side of geological research, and it is not an exaggeration to say that modern civilization is built on a foundation of resources discovered by geologists like him.[5] But he kept up his interest in pure research as well, studying the Apennines between voyages of exploration, and after his final return to Italy in 1927.

As early as 1901 Bonarelli wrote a detailed study of the rocks around Perugia,[6] which are very similar to those at Gubbio, although they extend farther back into time—back to the Jurassic. Most of the Jurassic is a complicated sequence of deep-water limestone and chert. Fortunately this complicated sequence of rocks is rich in ammonites, the coiled shells that are so good for finding the ages of rocks because of their rapid evolution and their floating all around the world in the surface waters of the seas. By 1901 geologists had devoted great effort to studying ammonites, and many different ammonite species could be used for accurate dating. As a result Bonarelli's dating of the Jurassic formations is quite close to the modern ones.[7]

But when we come to the Cretaceous *Scaglia* and younger rocks, there are huge errors in Bonarelli's age assignments. For example, he considered a formation called the *Scaglia cinerea* to be latest Cretaceous in age, when we now know it dates from the Oligocene. Using the ages in years that we now have, this means he thought a 25- to 30-million-year-old formation was about 65 to 75 million years old, an enormous error. How could such a good geologist have made such a mistake?

The problem was that Bonarelli had no fossils to rely on. It is clear from his writings that he searched very hard and just could not find fossils. The *Scaglia* simply does not contain ammonites or any of the other full-size fossils that geologists then used as their main dating tools.[8] His study was of high quality for the time, but accurate dating of the *Scaglia* would have to wait for a new generation of geologists. In chapter 10 we will see how dramatically the view of the Apennines changed when geologists learned to date rocks with tiny microfossils.

Returning to Gubbio to Search for Fossil Compasses

Our bleak and shivering introduction to Gubbio at Christmastime in 1970 was not a strong inducement to return. Yet we did come back three years later, in the fall, and Gubbio had changed completely. Warmed all summer by the Umbrian sun, the old stone buildings looked down with grace and dignity on streets and piazzas full of people, and on the little dramas of Italian life.

We found that the people of this town take pride in their nickname—*i matti di Gubbio*—the crazies of Gubbio—because of a wild race in which they carry three heavy wooden towers called *ceri*, crowned with statues of saints, at a dead run from the main piazza up a zigzag path to the Church of Sant'Ubaldo, high on the mountain above town. Like the Palio of Siena, the Festival of the Ceri each May totally absorbs the people of Gubbio.

Milly and I came back that fall of 1973, with Bill and Marcia Lowrie, to study the rocks exposed in the mountains behind Gubbio. Soon those rocks would totally absorb Bill's interest and mine, and in the years that followed, more and more geologists from all over the world would come to this place, as it became clear that the mountains of Gubbio hold one of the greatest historical archives ever assembled.

We went to Gubbio because the pink color of the *Scaglia rossa* suggests that it contains iron minerals rusted to a reddish hue, and iron minerals sometimes give rocks the ability to record the direction of the

The race of Ceri on the mountain above Gubbio in 1985.

Earth's magnetic field at the time they were formed. As a paleomagnetist, Bill can measure those ancient magnetic-field directions in his lab, as if the rocks contained fossil compasses.[9]

Bill and I hoped to find that the fossil compasses in the *Scaglia rossa* were rotated away from north, which would tell us about the possible rotation of Italy. Geologists had recently shown that the nearby islands of Corsica and Sardinia had rotated away from the southern coast of France, behaving as a "microplate."[10] Perhaps the Italian Peninsula was also a rotated microplate!

Our paleomagnetic approach worked, but not very well. The fossil compasses of the *Scaglia* pointed northwest—not north—indicating a counterclockwise rotation.[11] But we could not say very much about that rotation because the *Scaglia* limestone may have detached from the underlying rocks, and may have twisted around, so the fossil compasses in the *Scaglia* did not necessarily tell us the rotation history of the Italian crust far below, certainly not in any detail.

It was a distressing development, but we quickly forgot our disappointment when we discovered something much more exciting. Instead of a study of the local geology of the Mediterranean, we realized that we were looking at something of global significance.

We discovered, completely by accident, that the *Scaglia* bears a record of the reversals of the Earth's magnetic field! In some beds the fossil compasses point northwest, probably because of the rotation of the Italian crust, and in other beds they point the opposite direction.[12]

Geologists already knew about magnetic reversals, a really surprising discovery that was critical in the plate-tectonic revolution.[13] From time to time the Earth's magnetic field collapses and then comes back in the opposite direction. The daily spinning of our planet does not change, but after a reversal a compass that formerly pointed north would point south.

In the plate-tectonic revolution, geologists came to understand that continents like Africa and North America slowly move away from each other. New oceanic crust forms at the midocean ridge halfway between them, where deep mantle rises to fill the gap, in a process called sea-floor spreading. And as the rising hot mantle cools to form ocean crust, it gets magnetized, recording the direction of the Earth's magnetic field.

Meanwhile the Earth's magnetic field is reversing from time to time, and as a result, there are invisible stripes on the sea floor—stripes of normally and reversely magnetized sea floor, symmetrically arrayed on either side of the midocean ridge, and getting older away from the ridge. Those stripes can be detected by towing a magnetometer behind a ship or an airplane.

Bill and I had met as young researchers at the Lamont-Doherty Geological Observatory of Columbia University, near New York City, a hotbed of activity during the plate-tectonic revolution. When we first realized that reversals are recorded in the *Scaglia*, we immediately recognized the significance, because magnetic reversals were a hot topic at Lamont. Walter Pitman and his group were mapping magnetic stripes all over the world's oceans and using them to track the motions of continents as sea-floor spreading moved them around. Neil Opdyke, the paleomagnetist who had invited Bill to Lamont, had been able to date the youngest reversals back to just a few million years from deep-sea sediment cores. But there was no way to date the older reversals. Our Lamont friends could map the magnetic stripes, but they couldn't accurately determine the ages of the reversals they recorded. They could tell *where* the continents had moved, but they couldn't tell *when*.[14]

The record of reversals in the *Scaglia* promised a solution to that dilemma, because in the meantime paleontologists had solved the problem that had frustrated Guido Bonarelli. They now knew how to date the *Scaglia* in remarkable detail, using microfossils. Bill and I couldn't wait to get back to Italy to start measuring the magnetic reversals in the *Scaglia*. Gubbio could become the key to dating the history of continental drift all over the world!

Otto Renz and the Microfossil Heritage of Ambrogio Soldani

Ambrogio Soldani's pioneering work on microfossils at Siena in the late eighteenth century had to wait more than a hundred years before its potential was finally realized. But at last, at the beginning of the twentieth century, geologists began to make serious use of microfossils, and especially of the Foraminifera. With petroleum companies drilling wells all over the world in the search for oil, the tiny and often very abundant forams emerged as the ideal tool for determining the ages of marine sedimentary rocks.

As Soldani had done, specialists called micropaleontologists would wash away the soft matrix of clay and silt, clean off the forams, and study them under the microscope. Micropaleontologists named the genera and species of forams, worked out their evolutionary family tree, and tied that family tree into the standard geological timescale. Forams offered a wonderful new tool for dating rocks.

The *Scaglia* limestones of the Apennines are full of forams. Even Brother Elias must have seen them, as tiny specks, in the limestone from the Assisi quarries that he used to build the Basilica of Saint Francis.

But they were frustrating to micropaleontologists, because the *Scaglia* is so solid and hard that it is impossible to get the forams out whole. When you hit a piece of *Scaglia* with a hammer, it breaks right across the forams, so you cannot see the surface details of the little shell that were needed to identify the species.

The breakthrough came in the early 1930s. A young Swiss geology student named Otto Renz was studying at the University of Bologna, where he was encouraged to investigate the *Scaglia*. Italian geologists suspected that some of the Apennine rocks were much younger than Bonarelli had thought in 1901, and the problem needed a careful, detailed study. Otto Renz accepted the challenge, and from 1932 to 1934 he roamed through the Umbria-Marche Apennines on a bicycle, mapping the geology, measuring stratigraphic sections of the *Scaglia*, and collecting samples.[15]

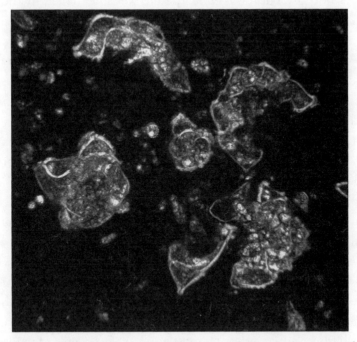

Foram shells seen with a microscope, in a thin section of the Scaglia *limestone. The foram in the upper right happens to be cut in a profile that makes it easy to identify, but the other cuts are not diagnostic.*

The best exposures Renz found were in the Bottaccione Gorge at Gubbio. He measured the beds, drew diagrams, and collected samples, just as we were to do forty years later. Back in the lab, he made thin sections of the *Scaglia*. This is a standard geological technique for studying rocks. You cut the rock with a diamond saw, polish the surface flat, glue it to a glass slide, and grind it down so thin that you can see through it. Then you can identify the minerals and study all the fine details under a microscope.

Otto Renz did something new with his thin sections—he used them to identify the forams in the *Scaglia*! It was tricky because micropaleontologists were used to determining the species of whole, cleaned-off specimens, as Soldani had done, while all Renz could see was random cuts through the forams. Most of the cuts were useless, but occasionally he would see a profile that was diagnostic, as when you see a recognizable silhouette of a person's head.

It was not easy, but it was enough. Renz was able to identify the forams in thin section well enough to recognize their basic evolutionary changes. Most important, he recognized that all the forams of the genus *Globotruncana* became extinct partway up through the *Scaglia*. This, he knew, marked the end of the Cretaceous, and it had two important consequences.

First it meant that Bonarelli in 1901 had been wrong in thinking that all of the *Scaglia*, up to the *Scaglia cinerea*, was Cretaceous in age. In fact the upper part of the *Scaglia rossa* is Paleocene and Eocene, and the *Scaglia cinerea* is Oligocene. The tiny microfossils made it possible to correct an error of more than 40 million years.

Second, when Renz recognized the extinction of the genus *Globotruncana* in the *Scaglia rossa*, he set in motion the chain of events that would lead, almost fifty years later, to the recognition of the impact in Mexico that caused the extinction of the dinosaurs.

As Otto Renz was finishing his study of the forams in the *Scaglia*, he returned to the Apennines on a field excursion, and he tells of the great pleasure he had in visiting the *Scaglia* outcrops at Gubbio with Guido Bonarelli, then in his sixties.[16] It must also have been a pleasure for Bonarelli to see how this young Swiss geologist could now use tiny fossils to date those hard limestones that had frustrated him when *he* was a young geologist.

Understanding and Using Forams

Ambrogio Soldani had no way of knowing what kind of little organisms made the tiny fossils he studied and drew for so long. Today's geologists would be able to use foram microfossils for dating and correlation even if we knew nothing about the organisms that made them, but the organisms are fascinating in themselves. Each foram may only be a single cell, but what that single cell can do is amazing.

First of all it builds itself a little shell by extracting calcium, carbon, and oxygen from the ocean water, precipitating it as the mineral calcite, and shaping it into ornate forms, sometimes of great beauty. The walls of the shell are full of tiny holes. This gives them their name—the word "Foraminifera" derives from a root that means "hole."

If all we had were fossils, we would not know what those tiny holes are for. But paleobiologists like Howie Spero at the University of California at Davis study living forams in the sea and in the laboratory. They find that forams extend long, thin threads of protoplasm from their single cell, out through the little holes, to make a web of hair-like filaments reaching out several times the diameter of the shell. The filaments are called pseudopoda, meaning "artificial legs," and they make a living foram beautiful to our eyes but fearsome and deadly to other organisms its own size.

The pseudopoda greatly extend the diameter of the foram, making a sticky web that entangles small animals of comparable size. In a photograph made by Howie Spero we can see an unfortunate little copepod that brushed against this deadly web. Such a victim probably exhausts itself in a vain struggle. Then the pseudopoda drag the copepod in toward the one large opening in the shell. Enzymes secreted by the pseudopoda decompose the victim, and the pieces are dragged in through the opening and digested in the main part of the cell.

Each foram makes a succession of larger and larger chambers to house the cell as it grows. Different sizes and shapes of walls, chambers, and openings have offered advantages to forams at different times in their evolution, and thus their evolutionary history has produced a whole variety of species that can be recognized by paleontologists and used to date the rocks that contain them.

Following in Otto Renz's footsteps, while she was still a young stu-

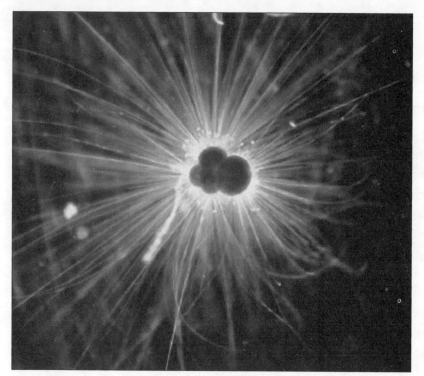

Howie Spero's microscope photograph shows the shell of a living foram in black. You can see several chambers and the array of filaments that the single-celled organism pushes out through little holes in the wall, to make a web surrounding the shell.

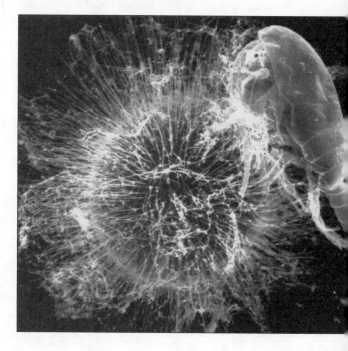

This remarkable electron microscope image by Howie Spero shows a foram that has trapped a little animal called a copepod in its sticky web, pulled it in close, and is about to begin digesting it. The foram and its victim would barely be visible to the naked eye.

dent in the 1960s, Isabella Premoli Silva of Milan made some major strides in using forams to understand the *Scaglia*. Isabella and Hans Peter Luterbacher made a careful study—at Gubbio, of course—and they were able to divide the *Scaglia rossa* into more zones than Renz had been able to recognize. This was important because it meant they could tell the ages of different parts of the *Scaglia* in more detail than in Renz's work.[17] Isabella has continued to refine the foram dating of the *Scaglia* throughout her career.[18] Her remarkable skill in recognizing forams in thin section made possible the work we did on dating the fingerprint of magnetic reversals in the *Scaglia* at Gubbio.

Magnetic Fingerprints

By 1974 Bill Lowrie and I had learned that the *Scaglia rossa* carries a record of reversals of the Earth's magnetic field. Did that sedimentary record of reversals match the record in the ocean crust? Would the *Scaglia* allow us to date the sea-floor magnetic stripes and thus date the history of continental drift? Bill and I had stumbled onto a really significant opportunity, so we set out to understand the *Scaglia* in detail.

As so often happens in science, we were not alone in making this discovery. At the very same time, an Italian American team led by my former professor at Princeton, Al Fischer; Florentine paleomagnetist Giovanni Napoleone; and Isabella Premoli Silva also discovered the record of magnetic reversals in the *Scaglia* at Gubbio. When we became aware of our simultaneous independent discovery, we decided to work together, and we all took great pleasure in the emerging understanding of the *Scaglia* and its historical record, and in our days of fieldwork together. We concentrated on Otto Renz's and Isabella's section in the Bottaccione Gorge in the mountains behind Gubbio, where a medieval aqueduct wanders along the mountainside, above a road that gave us easy access to the outcrops of *Scaglia rossa*.

The *Scaglia* limestone accumulated for a very long period of time and is almost 400 meters thick—like four football-field lengths piled on top of each other. In trying to tease out the record of magnetic reversals in this thick pile of limestone, we drilled many hundreds of little one-inch rock cores to measure in the lab, and we needed to know exactly where in the

The Bottaccione Gorge at Gubbio. The horizontal structure is the medieval aqueduct of Gubbio. We measured our stratigraphic section of Scaglia *limestone along the road, and there we took samples for our paleomagnetic study.*

pile of sedimentary beds each core came from. So we began by measuring a stratigraphic section.

Geologists have learned that measuring a section is the necessary starting point for almost any study of Earth history written in layered rocks. It is the practical application of Nicolaus Steno's seventeenth-century discovery that Earth history is written in sedimentary rocks, piled up bed on bed, from older to younger. As far as we know Steno never actually measured beds of rock, but he might very well have done so if he had stayed interested in geology.

At Rome, Albert Ammerman and I began our study of the Capitoline Hill by measuring sections in outcrop and in well cores. At Gubbio the task was harder, because the *Scaglia* is so much thicker than the volcanic layers in Rome.

Measuring the section would have been especially difficult if the *Scaglia* beds still had their original horizontal orientation, because four hundred meters of *Scaglia* would be like a quarter of the depth of the Grand Canyon, and we would have had to scale high cliffs to make our measurements. Fortunately the beds of *Scaglia* were tilted to about forty-five degrees when the Apennines were built. This made it possible to measure the entire section along the road that winds up through the Bottaccione Gorge. So Al Fischer and Mike Arthur patiently measured up through the beds of *Scaglia* with a meter stick, painting a mark every meter so that we could refer all our samples to a single measured section.

On our sampling campaigns Bill would drill out the cores with a one-inch, diamond-tipped steel tube driven by a noisy old chain-saw engine. Trying to ignore the angry snarling from the drill, I would find the position of each sample in our section—308.55 meters for example—and deter-

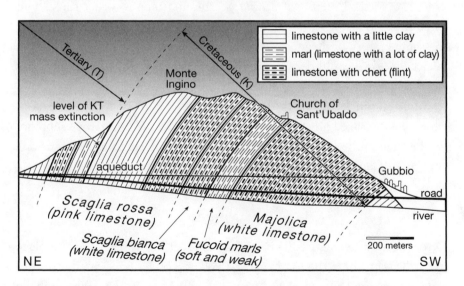

This diagram shows the south side of the Bottaccione Gorge, cutting through the mountains behind Gubbio. The beds were tilted during the building of the Apennines, which allowed us to measure the thickness of the Scaglia limestones by walking along the road.

mine the alignment of the core and the orientation of the limestone bed with a compass and a bubble level. Then I would carefully look at each core and make notes about the kind of rock—pure limestone, or limestone with some clay in it, pink or white in color, and anything else I could see with a little magnifying hand lens.

Gradually, over several years of fieldwork, Bill and I built up a record of the magnetic reversals in the Upper Cretaceous *Scaglia*. Giovanni Napoleone and Bill Roggenthen did the same in the Paleocene, while Isabella supplied the dates from her thin-section studies of forams. As the pattern of reversals emerged from the measurements, it became clearer and clearer that the pattern of long and short magnetic zones, normal and reversed, was an excellent match to the pattern in the ocean floor—it was like matching fingerprints or tree rings.

We came to understand that the Earth is running two different magnetic tape recorders, each of which records the reversals of the magnetic field. One recorder is the spreading sea floor, and the other is the accumulation of deep-sea limestones like the *Scaglia*, containing a tiny amount of magnetic iron minerals. The same pattern of reversals is recorded in about five hundred horizontal kilometers of ocean crust in the Atlantic Ocean as in 150 vertical meters of *Scaglia*, so the first tape recorder runs about three thousand times faster than the second, but they are both recording the same history of magnetic reversals. And because the *Scaglia* is full of microfossils, we could date the reversals not only in the *Scaglia*, but thus indirectly in the ocean floor as well.[19]

In 1976 our friend Paolo Pialli organized a conference at Perugia at which we discussed all the new results on the *Scaglia* at nearby Gubbio.[20] By the time of the Perugia conference, we had realized that there was one particular level in the *Scaglia rossa* that just cried out for explanation. . . .

Catastrophe

Otto Renz had recognized this intriguing level in the 1930s as the place in the *Scaglia* where the foram genus *Globotruncana* disappeared, never to return, marking the end of the Cretaceous Period. In the 1960s Isabella and Hans Peter Luterbacher had found this same marker for the end of the Cretaceous in twenty different locations in this part of the Apennines.[21]

At Gubbio during the 1976 Perugia conference. In the front row (left to right) are Roger Larson, Al Fischer, Walter Alvarez, Bill Roggenthen, and Paolo Pialli. Behind (left to right) are Roberto Colacicchi, Giovanni Napoleone, and Bill Lowrie. Missing from the picture are team members Isabella Premoli Silva and Mike Arthur. Slanting down diagonally from the upper right is a boundary between darker Tertiary Scaglia rossa above and lighter Cretaceous Scaglia rossa below. This change in the Scaglia would lead us to the next scientific adventure.

The end of the Cretaceous can easily be seen with the naked eye. At 347.6 meters in our measured section, the forams that characterize all the Cretaceous *Scaglia* suddenly disappear—or *almost* disappear. All the forams big enough to see with the naked eye were abruptly terminated. Two or three species survived into the base of the Paleogene—species so tiny that they can be seen only with a microscope. Those survivor species have evolved into all the forams we have today, but the forams just barely made it through this great extinction that geologists call the KT boundary.[22]

The KT boundary marks one of the six great mass extinctions documented in the fossil record. Not only forams, but many other kinds of plants and animals as well, either died out or nearly died out at that moment in Earth history. The coiled-shell ammonites and the great dinosaurs that walked the land were terminated forever.[23]

Up through the 1970s, most geologists believed that the KT mass extinction had been a slow, gradual dying off, because geologists had been trained to shun catastrophic explanations and to think that all Earth processes were slow and gradual—or "uniformitarian," to use the geological jargon.[24] But at Gubbio we could see that the near-extinction of forams took place across a little bed of clay only a half-inch thick. No one had ever had that kind of time resolution before, and it made one thing clear—the KT extinction had been very sudden, and not uniformitarian at all.[25] Something was wrong with the standard geological view of Earth history.

We now know that the KT mass extinction was the result of the impact of a huge comet or asteroid that hit Mexico's Yucatán Peninsula. This was the subject of my previous book, *T. rex and the Crater of Doom*, in 1997, and it was a remarkable detective story with contributions from hundreds of geologists all over the world.[26]

It began there in the Bottaccione Gorge at Gubbio in the summer of 1977, when Terry Engelder and I carefully collected a sample across the KT boundary.[27] It was not easy to collect, because there is a half-inch bed of soft clay separating the hard limestones of the top bed of Cretaceous *Scaglia* and the lowest bed of Tertiary *Scaglia*, and the clay kept breaking up and the limestone beds would come apart. But we wanted that clay in particular, because it represented exactly the time when the mass extinction occurred, and finally Terry and I quarried out a really good, coherent piece, which eventually became the key to the puzzle.

Back in Berkeley the next winter, our group discovered that the thin clay bed in our KT boundary sample from Gubbio has a strong enrichment in the rare element iridium, a fingerprint for extraterrestrial material. It took a couple of years to understand what that meant, but in 1980 we proposed that a giant impact had caused the KT extinction.[28]

A huge controversy ensued, for in science we find the answers to questions by intense debate, by searching for more evidence, and by challenging all evidence and all hypotheses. Some geologists argued strongly against the impact idea, while others found additional evidence to support it. In Spain, Jan Smit found droplets of rock that had been melted by the impact and thrown completely out of the atmosphere before falling back to Earth all around the planet. In Montana, Bruce Bohor found fallen sand grains made of the mineral quartz that showed unmistakable signs of a shock wave caused by the impact that had launched them out of the atmosphere. Like the iridium anomaly, the impact spherules and shocked quartz occurred exactly at the KT boundary and virtually nowhere else.[29]

The impact theory implied the existence of a huge crater, but the crater was well hidden—buried beneath a kilometer of Tertiary sediments, below the surface of the Yucatán Peninsula—and it was more than a decade before it came to light. Geologists of the Mexican Petroleum Company, PEMEX, had discovered it, and PEMEX geologists Antonio Camargo and Glen Penfield correctly interpreted it as a giant impact crater. But PEMEX kept it secret until tsunami deposits at the KT boundary in Texas led graduate student Alan Hildebrand to hunt for the crater in the nearby Gulf of Mexico. Eventually Alan heard about the studies by Camargo and Penfield, and they were finally allowed to talk. Now we know that this crater, centered beneath the fishing village of Puerto Chicxulub, is the largest impact crater yet found on Earth, and dates from exactly the moment of the KT mass extinction.[30]

The story of the KT impact and mass extinction is particularly dramatic, but it is just one of a host of discoveries about Earth history that have come from the archives that nature has written in the *Scaglia* limestone in the Apennines. Geologists keep finding new ways to extract historical information from these rocks, and there are always new geologists coming to Italy to study the *Scaglia* and the other pelagic limestones of the Mountains of Saint Francis.

In addition the *Scaglia* at Gubbio had yet one more way to help geologists with their critical task of finding the ages of rocks. It was not just dating sediments on the basis of microfossils and magnetic reversals that benefited from the rock record at Gubbio. These remarkable limestones also helped with the effort to attach numerical ages, in millions of years, to the geological timescale.

The Breakthrough to Ages in Years

We continued our work on magnetic reversals for several years, extending it down into the Fucoid marls and the *Majolica*, and up above the *Scaglia rossa* as well. One day in 1979 we were sampling the Oligocene *Scaglia cinerea* in a quarry near Gubbio. The *cinerea* is a soft marl—a limestone containing lots of clay—unlike the hard limestone of the *Scaglia rossa*. Bill was drilling, and the soft marl in the drill cores kept shattering. Milly followed along behind Bill, gluing the pieces of core together. I was measuring the orientations of the cores and looking carefully at each one with my hand lens. And then I saw something new. In one core there were little dark specks. Not forams, I could tell, but grains of mica.

This was a really exciting discovery, because geologists can determine ages in millions of years on mica grains. Fossils are great for tying rocks into the standard geological timescale—telling that a rock is Cretaceous or Oligocene in age—but they do not tell you the ages in years.

Mica contains potassium, which decays radioactively to argon at a known rate. By measuring the amount of potassium and of argon, you can determine how many million years it has taken for that much radioactive decay to take place. The more argon, the older the mineral grain. It's a simple concept, though difficult to put into practice, but geologists have become very skillful at getting really accurate ages in this way.[31]

However, if the mica grains in the *Scaglia cinerea* had been washed in by a river, their ages would have been much older than the *Scaglia*. So it was critical that there were abundant scatters of mica at just a few levels, and absolutely none anywhere else. This told us that the mica grains had been blasted into the sky by volcanic eruptions, and had been carried to this place by the wind. In terms of the volcanic rocks of chapter 3, the mica was deposited like air-fall tuff, but so far from the volcano that there was just a scattering of mica grains. The mica ages would tell the age of the eruption, which would be the same as the age of this level in the *Scaglia*. It was a perfect situation for getting ages in years on sedimentary rocks, which is normally very difficult to do.

At that time, in the early 1980s, I had a remarkable Italian graduate student named Sandro Montanari, and Sandro took on the challenge of dating the mica from the *Scaglia cinerea*. Sandro and a growing list of friends have kept at it for more than twenty years, gradually finding more and more levels of volcanic ash in the *Scaglia* and the overlying Miocene formations, and getting really accurate ages. This has been of global importance to geologists, because the dates in years from the Apennines, in sediments full of forams, have in turn provided high-quality ages on this part of the geological timescale.[32]

Sandro Montanari and Paula Metallo at the Geological Observatory of Coldigioco, a research and teaching center for reading the history written in the Apennine rocks.

So once again the *Scaglia* has been a big key to understanding Earth history. Why should this particular limestone contain such an amazing historical record? The answer is that the *Scaglia* is a very unusual limestone, deposited slowly on the deep-sea floor, far below the surface of the sea, as forams and other tiny grains slowly sink down through the water and come to rest on the bottom. Most limestone is deposited right at sea level, where storm waves frequently tear away portions of the just-deposited rock record, like vandals in a library, tearing pages out of the books. Storm waves do not reach the depths where the *Scaglia* accumulated, so in that quiet environment, every detail has been preserved.

In 1992, Sandro and his artist wife, Paula Metallo, moved from Berkeley back to Italy and founded the Geological Observatory of Coldigioco, not far from Gubbio.[33] Every year dozens of students and researchers come to Coldigioco, finding ever-new and different kinds of records to read from the Apennine rocks, which have turned out to be an almost inexhaustible historical archive.

The Timeline of Earth History

We now come to the geological timescale—the great chart that arranges everything we know about the history of the Earth into chronological order. Earth history has gone on for a very long time, and geologists have learned a great deal about what happened when, so the geological time-scale is unavoidably large and complicated. Anyone seeing the entire chart for the first time would be overwhelmed and unlikely to make the effort to understand it. But the geological timescale is a treasure chest of fascination, so let me lead you into the parts relevant to the Mountains of Saint Francis. This will serve as a framework for the rest of the book, in which we will explore some of the intellectual delights that lie below the complicated surface.

This first figure shows the entire timescale, for all of Earth history, but with only the coarsest divisions. It is a way of placing the history of the

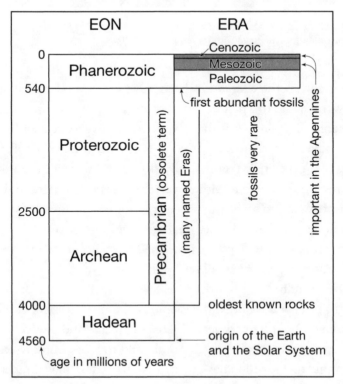

The geological timescale for all of Earth history, greatly simplified.

Mountains of Saint Francis in a broad context. The chart shows that abundant fossils are found only in the last tenth of all geological time.

The timescale for that last tenth of Earth history was constructed on the basis of fossils. Geologists were well aware of the law of faunal succession long before Darwin provided an explanation of why the shapes of fossils change as you work upward through the sequence of sedimentary rocks. Following in the footsteps of Nicolaus Steno, those early geologists collected lots of fossils and tried to make sense of what they found.

At the beginning of that effort, in the early nineteenth century, a geologist would go to a particular region and try to understand the local sequence of rocks. This led to a set of rather odd names for the major intervals of Earth history we now call periods. For example, the Triassic is named for a region in Germany where there were three prominent layers of rock, the Jurassic for the Jura Mountains of Switzerland, and the Cretaceous for parts of England where there is a lot of chalky limestone (*creta* means "chalk" in Latin). Tertiary remembers the very earliest attempt to subdivide geological time—into three intervals—by the eighteenth-century Italian geologist Giovanni Arduino.[34]

The story of the Mountains of Saint Francis deals with only the last half of the part of history in which fossils are abundant—with the eras called Mesozoic and Cenozoic, which mean "middle life" and "recent life."[35] Here, in a second figure, is a blowup of the part of the timescale we need for understanding the Apennines.

Later the geologists gradually assembled the local period names, in the proper order, into a global timescale. They found they could group the periods into larger time intervals that we now call eras.

A timescale with only a dozen or so subdivisions of Earth history would not have the resolution we need for a really detailed understanding of the past, so the periods have been divided into epochs, and those in turn into stages. And finally, within the stages, there are fossil zones, like that of the foram *Globotruncana calcarata*, based on first and last appearances, allowing paleontologists to say just where a rock fits into the timescale of Earth history.

Over the years geologists have defined and named a large number of epochs, stages, and fossil zones, and I doubt if anyone knows them all by heart. All geologists know the eras and periods, but we memorize only the part of the timescale we are working on. Studying the Apennines, I know the timescale for the Jurassic, Cretaceous, and Cenozoic, but not the older parts. If we need to refer to some other portion of the timescale, we can turn to books, charts, and Web sites that give the whole system.[36]

When you first look at the geological timescale, it seems just a Babel of unfamiliar, meaningless names—rather like walking into a room full of strangers. But some of the strangers may eventually turn into close friends, and as you get to know the timescale, the different periods and epochs and stages come to have unique personalities and remarkable tales to tell. Just as an example, one story (we will explore it in chapter 13) concerns

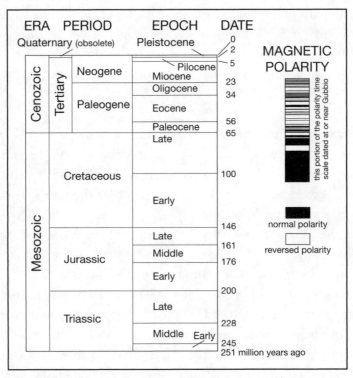

The portion of the timescale useful to geologists working in the Mountains of Saint Francis. I have shown the fossil-based eras, periods and epochs on the left, then the ages in millions of years, and then the portion of the magnetic polarity sequence dated at Gubbio.

the late Miocene stage called the Messinian—during which the entire Mediterranean Sea dried up, its floor becoming a baking-hot desert two miles below sea level!

And so, with the geological timescale, we see how all the efforts to construct the history of the Earth fit together. From Nicolaus Steno, starting it all off in Tuscany, through the geologists who used fossils and microfossils to make the first timescale, to the dating of the timescale by means of radioactive decay, to tying it to sea-floor spreading with magnetic reversals, to a variety of new techniques still being developed, the construction of the timescale has been a great scientific saga. And with the timescale as a basis, we are now ready to explore the history of the Apennines.

PART IV

The Apennines

*In which we explore the Apennine Mountains that make up the back-
bone of the Italian Peninsula, to investigate the third great question
of the book—How have the mountains and valleys of Earth's present
landscape come to be? We will see how an enormously powerful but
extremely slow Earth storm is creeping across Italy, gradually turning
the sediment of long ages into today's majestic mountains.*

7

FROM WINTER STORM TO EARTH STORM

MILLY AND I descended the steep, icy streets of Gubbio that afternoon in the winter of 1970 and departed, through a gate in the city walls, into the worsening weather. We were hoping to reach Ravenna, on the east coast, where we wanted to see the mosaics that date from the sixth century, when Ravenna and Rome were the Byzantine co-capitals of Italy.

According to an old road map we had, we could cross the Apennine watershed at a pass called Madonna della Cima, northeast of Gubbio. To get there we drove for the very first time up the road through the Bottaccione Gorge, past the outcrops of *Scaglia* that were soon to become so familiar. But the pass was closed by snow, and our map showed no other obvious route through the mountains.

Back at Gubbio we asked directions of someone who was amused by our antiquated map and who pointed us to a new road up the neighboring Contessa Valley. We crossed the divide through a smoky tunnel and had our first view of the high Apennines. All we could see was a massive limestone ridge, mostly lost in the cloud deck. The snow-covered ridge appeared to block the way eastward.

It was our first glimpse of the Umbria-Marche Apennines—the moun-

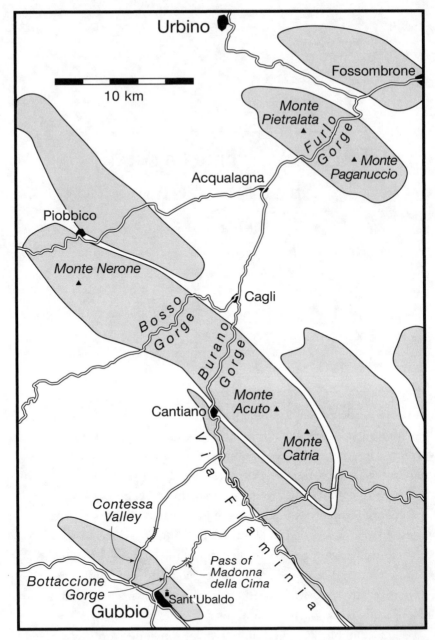

The gray areas show the big limestone ridges along the road from Gubbio to Urbino. The Contessa Valley road tunnels through the Apennine watershed. It then joins the Via Flaminia that cuts through the great limestone ridge, using the Burano Gorge between Cantiano and Cagli. The Flaminia then passes through the historic Furlo Gorge, and a side road leads west to the artistic city of Urbino. This map shows the roads before construction in the 1980s and 1990s converted parts of the Via Flaminia into a major highway. Today there is a tunnel beneath Monte Pietralata, so most drivers no longer see the Furlo Gorge.

tains whose history will be the focus of the rest of this book, providing an opportunity to understand how geologists recover the past of our planet.

Soon our route joined the Via Flaminia—one of the ancient roads that radiate out from Rome. The Flaminia played a key role when Goths and Byzantines were fighting to control Italy in the sixth century, because it linked the two Late Roman capitals, Rome and Ravenna.

Reaching the massive ridge, the Via Flaminia entered a rocky canyon called the Burano Gorge, winding along above an icy river. Limestone crags, dappled with snow, rose up and disappeared into the clouds. At the end of the Burano Gorge we emerged into the lowlands on the other side of the great ridge, at the little city of Cagli. The weather was getting worse. The light was fading, fog completely blocked the view, and a slushy rain began to fall. Our old road map showed nothing about the topography, and we guessed it would be low hills and plains all the way down to the coast. We stopped trying to see the landscape and paid attention to the dangerous driving conditions along the barely modernized Roman Via Flaminia.

Thus it was a shock when a momentary break in the fog revealed that we were not in the flatlands at all. Jagged cliffs rose on the left, and to the right a canyon fell off into the mist. We pulled over to look at the map, which gave no indication of any mountains here at all. I took a quick photograph so I would not think I had just imagined it.

In a few moments the fog closed in, and we drove on to Urbino, steeped in its memories of the Renaissance painter Raphael. We didn't even know that the name of the mystery canyon was Furlo, and, as with Gubbio, we could not have guessed how important this place would become to us.

"Tectonics"—the Architecture of Mountains

Several years passed. Bill Lowrie and I were working with Al Fischer's group, reading the history of magnetic reversals recorded in the *Scaglia* at Gubbio, and at Berkeley we were trying to understand the mass extinction that had brought the Cretaceous to a close. In addition I had started trying to understand how the Mountains of Saint Francis had come to be. What forces in the distant past had pushed up the massive ridges of the high Apennines? This required a change in mental focus. Rocks like the

Scaglia that serve as historical archives of reversals and extinctions also play a completely different geological role—as the building blocks from which the mountain range was constructed.

Gubbio lies some distance from the high Apennines and was no longer a convenient base for the new project. So we moved our headquarters to Piobbico, at the foot of a mountain called Monte Nerone. This peak lies at the northwest end of the great limestone ridge that is sliced through by the Burano Valley.

Over the years Milly and I spent many months in Piobbico, going out each day to map, measure, photograph, and sample the rocks of the Apennines. Piobbico, clustered around a medieval and Renaissance castle, has a slightly quixotic charm. It is known throughout Italy as the headquarters of the "Club of the Uglies," whose motto declares that to be *brutto*, or ugly, is a virtue, but to be *bello*, or beautiful, is slavery. Led for many years by Telesforo Iacobelli, who owned the general store, the club celebrates, with a light touch, the dignity and little-appreciated advantages of being unbeautiful, and each year awards the coveted "No-bel Prize."

The mystery canyon in the snow. We later learned that it is called Furlo and that it is rich in history, of both the human and the geological kinds.

In Piobbico I developed a close friendship with Father Domenico Rinaldini, affectionately known as Don Dò to his parishioners. Don Domenico was a classic mountain priest. He was short, broad, and very strong. He loved Monte Nerone, and whenever he had a chance, he would be up on the mountain, collecting the coiled-shell ammonite fossils that are found in the Jurassic limestones of the Apennines.

I remember one such day I spent with Don Domenico. We were up on the mountain examining the beds in a quarry of Jurassic limestones I was studying together with a fine group of ammonite paleontologists from the University of Rome—Professor Giovanni Pallini and his former students, who had become his close colleagues.[1] Whenever the Pallini group came to Piobbico, they would check in with Don Domenico to see his latest ammonites, and occasionally he would have a new species that they would describe and name. The Jurassic is best known as a time of dinosaurs, but of course there are no dinosaur fossils in these marine rocks. However, the marine ammonites are more useful to geologists than the dinosaurs are, for the ammonites evolved very rapidly, and their changing shapes make it possible to date Jurassic sediments in great detail.

At first it seems illogical to climb a mountain in order to look at the Jurassic. The Jurassic rocks are the oldest ones you can see exposed anywhere in this part of the Apennines. Jurassic layers usually lie buried deep beneath all the younger Apennine rocks. So why would you climb a mountain to look at the oldest, most deeply buried rocks?

The surprising answer is that Monte Nerone lies on the crest of an *anticline*—a great upfold of the strata.[2] Indeed the Jurassic *does* lie deep down in many places nearby. But at Monte Nerone the Jurassic, along with the other rocks above and below it, has been bent upward into a great arch. Rocks raised to high elevations are removed by erosion. On Monte Nerone erosion has stripped off all the Cenozoic and has eaten away the Cretaceous in many places, exposing the uplifted Jurassic.

For a stratigrapher, looking for historical information written in rocks, the anticlines provide good opportunities to see the oldest rock layers. In addition, for geologists interested in the mountains themselves, the anticlines pose a new puzzle: How and why did the Jurassic rocks get pushed up so high? And how did the whole mountain range come to be? Geologists call the large-scale study of mountain belts "tectonics," and to begin

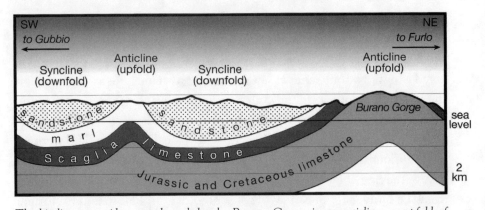

The big limestone ridge, cut through by the Burano Gorge, is an anticline, or upfold of the layers of rock. To the southwest are two synclines, or downfolds, separated by another anticline. Because of this folding, the Jurassic limestone is exposed in the big anticline but buried deep beneath the synclines. This cross section is the first of series in which we gradually explore the deep structure that underlies the Mountains of Saint Francis. Here we see the folds as they would have been interpreted by a geologist of several decades ago, who had only the surface geology to go on and did not have our modern understanding of how folds are produced. Further drawings that develop an understanding of this cross section are given on pages 196 and 214.

a tectonic study you need a clear picture of the geography of the mountain belt you hope to understand.

The View from Monte Nerone

Monte Nerone affords a view out over all the nearby landscape—the view I described in the Prelude. On that clear morning Don Domenico and I could see far across the belt of mountains. Peaks that had remained hidden inside the deck of storm clouds when Milly and I first crossed the Apennines now stood out sharp and clear in the morning sunshine. We could see the whole pattern of ridges and peaks, valleys and canyons.

Monte Nerone itself lies at the northwest end of a long, slightly curving anticline, shaped something like a banana sliced lengthwise and laid flat slide down. At the southeast end are the two peaks of Monte Acuto and Monte Catria, rising slightly higher than Monte Nerone. The middle part of the anticline is lower and is sliced through by two great river canyons—the Bosso Gorge and the Burano Gorge, where the Via Flaminia cuts through the barrier presented by the anticlinal ridge.

Looking southeastward from Monte Nerone toward Monte Acuto and Monte Catria, we could see the profile of the anticlinal upfold. It is very obvious, because of the Fucoid marls, dating from the Middle Cretaceous—the same as we saw at Gubbio. Marls are a soft mixture of clay and limestone that offers little resistance to erosion. As a result, the Fucoid marls have been stripped off the top of the resistant Lower Cretaceous limestone beneath it. The cleaned-off upper surface of the Lower Cretaceous *Majolica* limestone curves upward like a barrel vault, showing the arching structure of the anticline.

Off to the northeast Don Domenico and I could see another big anticlinal ridge, this one also sliced through by a gorge. It was Furlo, the mystery canyon once hidden in mist and snow, now fully in view. Don Domenico pointed out the tiny village of Pietralata, on the flank of the ridge, where he had rebuilt the church with his own hands as a young priest in the dark

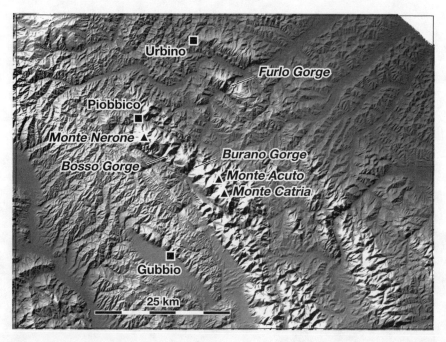

This topographic map clearly shows the ridge of the big anticline that extends from Monte Nerone southeast to Monte Catria. The anticline is cut through by the Burano and Bosso gorges, and a slightly smaller anticline to the northeast is cut by the Furlo Gorge. Geological structures like these folds are reflected in the landscape because they are young and have not yet been eroded away.

Looking northwest toward Monte Nerone, across the Bosso Valley. The arching profile of the anticline, with the beds bent upward, is very clear in the mountainside on the other side of the valley. The lower half of the mountainside in the foreground is made of Jurassic formations from the Corniola up to the Diaspri. The upper half, striped by bands of vegetation, is the Lower Cretaceous Majolica. The weak Middle Cretaceous Fucoid marls have been erosionally stripped off the top of the more resistant Majolica to give the gently arched erosion surface. The Upper Cretaceous Scaglia has been removed over most of the anticline, but a remnant is preserved in the dark, wooded mountain in the right-hand third of the picture.

The view northeast from Monte Nerone. On the skyline is the anticlinal ridge cut through by the Furlo Gorge, where once Byzantines and Goths battled for control of the ancient Roman road called the Via Flaminia. The gorge is guarded by Monte Pietralata on the left and Monte Paganuccio on the right.

days during and just after World War II. Pietralata was to become another of our wonderful stratigraphic sections for the paleomagnetic study of the *Scaglia*.[3]

The view of the Apennines from Monte Nerone really makes you puzzle over the origin of mountains. What forces bent up those anticlinal ridges? What Earth processes can lift rocks high up into the air? It is a fascinating but complicated story—one that will require the rest of the chapters in this book to bring to life.

One thing is clear from the start: The building of the Apennines happened very slowly, over immense periods of time. During this long interval, a dramatic convulsion has swept across Italy, from west to east, bending and breaking rocks and driving them over each other. It is not too far-fetched to think of this wave of deformation as a great Earth storm, but a storm moving so slowly that it is imperceptible from the human perspective. To understand mountains we must first try to appreciate these very slow changes over very long intervals of time.

Earth Storms Too Slow to Perceive

Few people other than geologists are familiar with the concept that a mountain range is born, grows, and finally disappears. It takes some getting used to the idea that a majestic range of snowy peaks comes into being where previously there were no mountains, and passes through a sequence of stages until it is finally worn away by erosion, looking very different at different times in its history. The reason for this unfamiliarity is of course the usual problem with Earth history—that it has gone on a million times longer than human history, with our lives and experience being so fleeting compared to the majestic unfolding of the life cycle of a mountain range.

The people of an Alpine village see the same unchanged mountain landscape in their old age that they remember from their childhood, unless there has been a big landslide, a volcanic eruption, or a glacial retreat. To the marine animals living on the sea bottom where a mountain range is in its earliest stages, the sea floor must seem fixed and eternally unchanging.

But if we speed up the passage of time a millionfold, the changes would be very noticeable in a single human lifetime. We would see a wave of deformation sweeping across the sea floor, lifting it above the surface of the waters, breaking and folding the rocks and raising them far into the sky. No sooner are the rocks raised up than they are torn apart and carried away by rivers and glaciers.

The Apennines began to form about thirty million years ago. If we lived in million-years instead of years, we would see the entire history of the Apennines in half a normal lifetime, and instead of the mountains being a permanent constant in our lives, they would be a dynamic feature whose changes would be as inescapable as the growth of children and the aging of our friends.

Suppose there were an advanced extraterrestrial civilization, and those beings lived very long lives. And suppose they had developed a particular interest in the evolution of the Earth. Suppose they sent a probe to visit our planet every million years, photographing and recording everything they saw on each visit. In their archives they would have a record not only of the evolution of life on Earth, but the evolution of its mountains as well, and the changing shapes of its oceans and the shifting positions of the continents.

Geologists have the job of constructing that archive. In some ways the extraterrestrials would have an advantage, for their files would contain photographs of Earth's surface showing the peaks and canyons of mountain ranges that have been completely lost to erosion. But frankly I prefer the way we do it as geologists, for we are not limited to taking pictures of Earth's surface, like the imaginary extraterrestrials. We can study the eroded roots of the mountains and thus understand what was going on deep beneath the surface throughout the entire history of a mountain range. We may not be able to say exactly what the peaks and canyons of long-vanished mountains looked like at each stage in their history, but we can understand why they formed, what forces brought them into being, and how the complicated deformations and displacements at depth produced the mountains we see today.

It is like the difference between the superficial history we learn in high school—just a catalog of names, dates, and events—and the intellectually more satisfying history at a college level that explores the fundamental trends and underlying causes of what happened in human history.

Having absorbed the scale change from years to million-years, geologists find the evolution of mountain ranges fascinating. Milly and I first caught a glimpse of the cliffs of the Furlo Gorge during a winter storm that was over and gone in a couple of days, leaving no lasting remains behind. The anticline of Furlo itself was built during an *Earth* storm that swept across Italy, at a rate so slow as to be imperceptible to us humans. But the Earth storm left a record of its passage, and that record is the Apennines themselves.

So let us explore the Apennines, studying the evidence preserved in the rocks of the cliffs and canyons. Let us put together an archive of their history, so we can watch as their sped-up evolution unfolds before us, giving an understanding of how these mountains were constructed.

8

Rocks for Building
a Mountain Range

I REMEMBER one day when Bill Lowrie and I were just getting acquainted
with the Mountains of Saint Francis and surprises awaited us everywhere.
We were driving from Cagli toward the Bosso Gorge, looking for good
rock outcrops for paleomagnetic studies. Around a bend in the road a
view opened up, and high on the mountain ahead there was a tall cliff.
With binoculars we could see that the beds of rock in the cliff were crumpled into tight, angular folds, different from anything we had seen before.
We did not yet know the Apennines well enough to recognize what rock
unit had been folded in this remarkable way.

We found a dirt road, half washed out, that climbed up to the cliff at
a place called Poggio le Guaine, and when we got there we realized that
the folded beds corresponded to our friend from Gubbio, the *Scaglia bianca*
limestone. Lower down in the cliff we could see the weaker beds of the
Fucoid marls, just as in the Bottaccione Gorge (page 103).

It brought home to us in a dramatic way that the Apennine limestone
beds represent more than just a library that carries a record of the times
when the beds were deposited. They are also the building blocks from
which the mountain range has been constructed. These cliffs interested us

The cliffs at Poggio le Guaine, where the strong beds of Scaglia bianca *limestone above, and the weak* Fucoid *marls below, have been crumpled into angular "kink folds." Here Bill Lowrie and I realized that the Apennine limestones serve not only as archives of Earth history but also as the construction materials with which the mountains were built. The cliff is about fifty meters (about 150 feet) high.*

in both ways, and we ended up studying both the paleomagnetic historical record in the rocks and the way the beds were deformed into kink folds, during the building of the Apennines.[1]

So before we trace just how the Apennines were built, let us take this chapter to understand the whole range of rocks that went into their construction.

A Threefold Arrangement of Rocks

Back in the seventeenth and eighteenth centuries, the earliest geologists struggled to understand the amazing variety of rocks they came across in the hills and mountains. At first many kinds of rocks made no sense whatsoever. Gradually, as they figured out the origins of these different mystery rocks, geologists found it useful to arrange rocks into three great groupings—metamorphic, igneous, and sedimentary.[2] Sedimentary rocks are dominant in the Mountains of Saint Francis, but the other two kinds play important roles as well. Let us see briefly what each of these three names means.

The name "metamorphic" means "changed in form," and the changes take place far below the Earth's surface. Deep down in the roots of growing mountain ranges, temperatures and pressures get so high that rocks can flow, deforming into twisted, contorted folds, while the mineral grains react chemically to form new minerals.

The Apennines do not offer many opportunities to study metamorphic geology, for only in a few places has erosion had time to expose the roots of these young mountains. The Alpi Apuane—the Apuan Alps of Tuscany—offer one spectacular exception, with quarries cutting into marble that formed where limestones were cooked and distorted under deep metamorphic conditions. Pure white Apuan marble provided the material from which Michelangelo and Bernini made their wonderful sculptures. The pure white marble is beautiful, but for a geologist, it tells a less interesting story than does marble with

bands and layers deformed into intricate folds. My first graduate student, Roy Kligfield, studied the folded Apuan marbles for his PhD thesis, and Roy's photographs of marble quarry faces convey a sense of the intense metamorphic deformation that can take place in the roots of mountain ranges.[3]

Temperatures deep within a growing mountain belt may get so high that the metamorphic rocks actually melt. Geologists

Roy Kligfield and the folded marble of the Apuan Alps in Tuscany. Quarrymen cut this beautiful rock face that shows how originally flat beds can be intensely deformed in the deep, hot, metamorphic roots of a mountain range, far below the Earth's surface.

call the molten, liquid rock "magma," and any rock that cooled from a magma they place in the second of the three great rock classes—the igneous rocks. The root of the word means "fire," as in "ignite."

Igneous rocks that form down deep, insulated by a great thickness of overlying rock, cool so slowly that the mineral grains have time to grow to large sizes. Coarse-grained igneous rocks, such as granites, thus bear witness to a deep origin, and geologists call them plutonic rocks, remembering Pluto, the god of the underworld. Like the metamorphic rocks, plutonic igneous rocks are rare in the Apennines, a mountain belt so young that almost nowhere has erosion exposed the deep levels where granites abound.

But sometimes magmas do not remain at great depth. Being hot, they rise buoyantly, and being fluid, they can squeeze upward through cracks and crevices until the magma reaches the Earth's surface. Magma erupting on the surface builds volcanoes, and geologists call these igneous rocks volcanic. Cooling rapidly at the Earth's surface, the minerals do not have time to grow to large dimensions, so volcanic rocks can be recognized by their fine grain size. These were the kind of rocks I had already studied in the hills around Rome.

By the time Milly and I went to Gubbio with Bill and Marcia, I had acquired some experience with all the major types of rocks—except for sedimentary rocks—the ones that slowly build up at the Earth's surface by the accumulation of various kinds of debris. The very young Apennines have seen so little erosion that they expose almost nothing but sedimentary rocks, and that is all you can see at the surface at Gubbio. But deep down beneath Gubbio, out of sight, there are ancient metamorphic and igneous rocks that played an essential role in the origin of the Apennines, although no geologist has ever seen them.

A Much Older Mountain Range, Buried Beneath the Apennines

Down deep, hidden from view, is an enormous body of rock called the Italian continental crust, going down to twenty kilometers in some places, forty in others, below the Earth's surface (twelve to twenty-five miles). Its top is the floor on which all the Apennine sedimentary rocks rest.

We know it is down there because we can see its top on seismic profiles, as discussed in chapter 11. A few oil exploration wells have reached the top of the continental crust in nearby areas, and it is exposed to view in a few places in Tuscany. From this scanty information we know in a general way the nature of the Italian continental crust, and it comes at first as a surprise. Its top is the eroded surface of another mountain range, much older than the Apennines, worn down almost flat by erosion.

This Italian crust is part of a very broad region in southern Europe that was mountainous in the Late Paleozoic, around three hundred million years ago. Geologists call these mountains the Hercynian Chain, and although they are buried throughout most of Italy, the metamorphic and plutonic rocks of the eroded-off roots of the Hercynian Mountains are exposed across much of Spain and France, and on the islands of Corsica and Sardinia.[4]

The top of the Italian continental crust is a fundamental geological break. Geologists call the old crust the "basement." It might have been better to call it the "floor," because it acts like the floor of a house when a rug gets pushed across it, crumpling up into folds—a good analogy to the way the Apennine anticlinal folds were built.

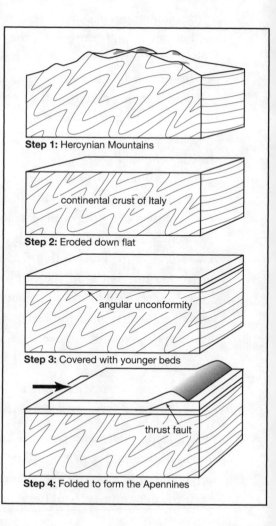

Step 1: Hercynian Mountains

Step 2: Eroded down flat
continental crust of Italy

Step 3: Covered with younger beds
angular unconformity

Step 4: Folded to form the Apennines
thrust fault

These diagrams show, in a very simplified way, how the ancient Hercynian mountains of Late Paleozoic time were eroded down flat, covered with younger sediments in Mesozoic and Cenozoic times, and how the thrust faulting of the younger sediments gave rise to the Apennine folds.

The basement is covered by a thick sequence of sedimentary rocks dating from the Upper Triassic—also nowhere exposed to view—but we do have some information about this Triassic formation. The information comes from a few oil exploration wells drilled in the Apennines after World War II, because the Italian petroleum company, AGIP, was hoping to find oil in these mountains. Oil often gets trapped in anticlines, so AGIP drilled a deep well in the Burano Gorge, right on the axis of the big anticline.[5] Unfortunately for AGIP but fortunately for the charm of the Apennines, there were no signs of petroleum, so after a few more test wells, this part of Italy was crossed off the list of potential oil provinces.

AGIP and its geologists have graciously made the information from those wells public, so we know something about the rocks of the Apennines that are too deep to see.[6] The Burano well did not reach the basement, but the drillers found a thick sequence of sedimentary rocks dating from the Upper Triassic. Because it was found in the Burano Well, in the Burano Gorge, the AGIP geologists named this hidden rock sequence the Burano Formation. This formation is rich in a mineral called anhydrite, which forms by evaporation of seawater in places where the water is very shallow and the climate dry and hot, which must have been the situation after the planed-off roots of the Hercynian Mountains were first submerged by the sea.

The Burano anhydrite is extremely important for understanding the origin of the Apennines, because minerals that form by evaporation are weak and easily deformed. The Burano acts like a layer of grease beneath the pile of sedimentary rocks. In many places the formations above it have come loose from the deeper levels and been have pushed up into a set of folds, just like a rug pushed across the floor. As we go on we will see that the Burano is the lowest of three layers of "grease," where detachment is easy, and these weak layers have played a critical role in the way the Apennine Mountains have evolved.

Earth Cycles, from Dante to Plate Tectonics

This glimpse of the ancient, forgotten Hercynian mountain range, leveled by erosion and buried under younger rocks, brings into focus one of the most remarkable discoveries geologists have made about the Earth—

the discovery that much of Earth history has unfolded in great cyclical patterns. The recognition of these slow, stately cycles in the evolution of the Earth gives us a profound insight into the past of our planet.

Way back in the fourteenth century Dante Alighieri understood one of these geological cycles. He realized that water evaporates from the sea, making clouds that give rain, and that the rainwater flows back down the rivers to the sea. Dante described this water cycle in a poetic manner when talking of the river Arno in the *Divine Comedy*,[7] and this part of his geological view has held up to scrutiny better than his theory that the deep Earth was the site of an "Inferno" full of suffering sinners.

In the nineteenth century, geologists made the more profound discovery that *rocks* also go through cycles. The rock cycle is very much slower than the water cycle, of course, and takes place over so many human lifetimes that this was a very difficult thing to discover. The two cycles are closely related, however, because the water cycle, keeping the rivers running, is responsible for the continuous erosion of the land and the transport of the erosional debris down to lower areas, which is part of the rock cycle.

By studying rocks of all kinds, those early geologists came to understand that the eroded sediments accumulate in low areas on the Earth's surface, getting buried deeper and deeper and solidifying into sedimentary rocks. If burial continues the sedimentary rocks may be squeezed and cooked into metamorphic rocks, or even melted and then cooled to make igneous rocks. Some of the melted rock may rise to the Earth's surface and erupt from volcanoes. And even the deep rocks may eventually be uplifted and eroded, with the debris from the erosion going to make the next generation of sediments. Perhaps the term "cycle" gives too simple a picture. The rock cycle is not a single loop, but rather a whole variety of looping pathways rocks can follow as they change from one form to another.

The rock cycle was an amazing discovery about the Earth, and ever since then geologists have devoted much of their effort to tracing the history of rocks all over the Earth as they have slowly moved through the various pathways of the rock cycle.

And yet, in the late twentieth century, geologists came to realize that the rock cycle was but a small manifestation of a still greater Earth cycle. In the late 1960s and early 1970s a flood of new evidence convinced geologists that continents move around on the surface of the Earth, and this

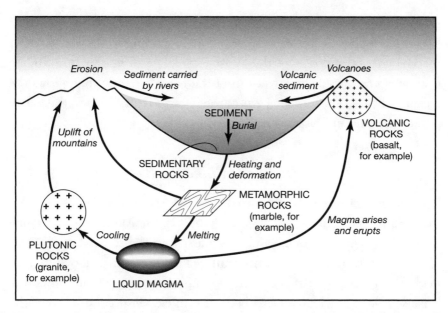

The rock cycle, one of the great discoveries of geology, shows the various pathways by which rocks of one kind are transformed into another. Although called a cycle, it is really more of a network of possible changes, and it is fundamental to understanding how the Earth works over long periods of time.

new view is called the theory of plate tectonics. The evidence showed that sometimes in Earth history there are several separate continents—each with its own continental crust, like that which lies beneath the Apennines. At other times all the continental crust is gathered together in a supercontinent. Today we are in a time with several continents, but 250 million years ago there was a single enormous supercontinent that geologists call Pangaea. The cycle from supercontinent to several dispersed continents and back to a new supercontinent is called the Wilson cycle.[8]

When a supercontinent breaks apart, hot mantle rises into the widening crack, and molten magma pours out of volcanoes. When two continents collide, the rocks in the collision zone may be buried, deformed, and cooked into metamorphic rocks or melted to make granites, and then uplifted and eroded. The buried Hercynian Mountains of the Italian continental crust and the Apennine Mountains that grew on top of the crust represent the adventures of continents caught up in the Wilson cycle.

Most of the exciting parts of the rock cycle happen either during continental breakup or continental collision, so it is the continental cycle that drives the rock cycle. And now we are coming to understand that the continental motions of the Wilson cycle are driven, in turn, by an even greater cycle of slow overturn of the Earth's mantle that allows the escape of heat from the deep parts of the planet. To understand the Earth, a geologist must always be thinking about cycles.

When Italy Looked Like the Bahamas

The oldest rocks exposed in our part of the Apennines are white limestones that the Italian geologists call the *Calcare Massiccio*. The name means "massive limestone," and it is massive in the sense that the beds are very thick, in contrast to the thin-bedded limestones above it. This made the *Massiccio* critical in the folding of the Apennine anticlines, because it was much harder to bend the thick *Massiccio* beds than the thin-bedded limestones above it. It is easier to fold a deck of cards than to fold a brick!

Because of the thick, strong beds, the *Massiccio* is resistant to erosion, and it can be recognized anywhere in the Mountains of Saint Francis by its great white cliffs. If enough *Massiccio* is exposed, the cliffs can be very high, for the limestone is seven hundred meters (almost half a mile) in thickness.

The rocks in the Furlo Gorge that Milly and I saw dimly through the clouds during the winter storm are *Calcare Massiccio*. On a sunny summer day the cliffs at Furlo are dramatic, and they can give us a gut feeling for the importance of this single rock formation as a key material in the construction of the Apennines.

Geologists are always interested in the environment where particular rocks accumulated, and the rocks often bear evidence of how and where they were deposited. Break off a piece of *Massiccio* limestone, wet the fresh surface, and look at it with a hand lens, and you will see a myriad of tiny white pellets. These are fecal pellets, shaped by animals that lived on a shallow sea floor, ate the chalky white carbonate mud of the sea bottom, digested any organic matter it contained, and excreted the rest as pellets.[9] If we needed yet another demonstration of the enormous extent of geologic time, we could look at the half-mile thickness of *Massiccio*, forming great cliffs, and deposited pellet by fecal pellet, by tiny organisms, in a small fraction of the Jurassic Period alone!

What was the sea floor like when the *Massiccio* was being deposited? The limestone contains no sand grains and very little clay, so it must have been far from the mouth of any river. Early Jurassic limestone like the *Massiccio* is found all over the Italian Peninsula, the southern Alps, and the countries of the former Yugoslavia, so this very-shallow-water environment covered a large area. Is there any place like that

The cliffs of Calcare Massiccio limestone in the Furlo Gorge are three hundred to four hundred meters high (about one thousand feet). This limestone was deposited on a shallow sea floor like that surrounding the Bahamas today. The great thickness and the lack of bedding planes made it into a very resistant unit that controlled the shape of the Apennine folds.

today? The answer came from Bruno D'Argenio, the first Italian geologist I ever met, when he was a postdoctoral researcher working with Al Fischer at Prince-ton and I was a graduate student. Bruno pointed out that the *Massiccio* sea floor in the Jurassic was like the present-day sea floor around the Bahamas and Florida, where pelletal limestones are accumulating today.[10] The Bahamas are small islands, slightly above sea level, within a broad platform slightly below sea level. At the end of this chapter we will see from Bruno's further work that the analogy between the *Massiccio* and the Bahamas is even more precise, and is critical for understanding the origin of the Alps and the Apennines.

Let us pause for a moment and note that limestones like the *Massiccio* are interesting to human beings for a more fundamental reason than just their role in the origin of the Apennines. Limestones have been absolutely essential in making the Earth habitable for plants and animals. What lies behind this unexpected statement?

Limestones are accumulations of grains of the mineral calcite, which has the chemical formula $CaCO_3$. If calcite breaks down, it releases carbon dioxide gas, whose composition, CO_2, can be seen in the calcite formula, along with a calcium atom, Ca, and another oxygen atom. If we think of the formula of calcite as $CaO + CO_2$, we can understand that calcite and limestone are nature's way of storing carbon dioxide gas in a solid rock, which keeps it out of the atmosphere.

Today, early in the twenty-first century, scientists and policy makers are seriously concerned about very small increases in atmospheric CO_2 from our human activities. CO_2 gas can trap excess solar heat and cause global greenhouse warming, with serious consequences for global society. But colossal amounts of CO_2 gas have been safely stored by nature in limestones. If it were in the atmosphere instead, this gas would trap so much heat that Earth would be as lethally hot as Venus, where there are no limestones to keep CO_2 out of the atmosphere. We living creatures owe a great debt of gratitude to limestones like the *Calcare Massiccio*!

An Ancient Atlantis, Sunk Beneath the Sea

The rocks *above* the *Calcare Massiccio* are abundantly exposed to our view throughout the Mountains of Saint Francis. They are also limestones, but the bedding is different, for the massive beds that make up the *Massiccio*, each of them a meter or more thick, give way upward to much thinner beds, several to the meter. By measuring the thickness of the formations and comparing them to the time intervals during which the formations were deposited, we can see that the rate of sedimentation slowed dramatically in the rocks above the *Massiccio*. The texture of the rock, visible with a hand lens, also changes, for the pellets of the *Massiccio* are missing from the formations above it.

The thinner beds, the slower sedimentation rate, and the absence of pellets signal a major change in the environment where the Apennine limestones were deposited. No longer was this part of Italy a shallow, Bahamas-style sea, bathed in sunlight, with photosynthesis producing a rich food source and lots of calcite mud to make the limestone.

The character of the thin-bedded formations above the *Massiccio* indicates that they were deposited in deeper water. Somehow the sea floor sank and the conditions that saw the deposition of the *Massiccio* changed to the setting of the deeper-water sediments that geologists call pelagic limestones, from the Greek word *pelagos*, for "sea."

At first one thinks of the sea floor suddenly sinking, but the change was more subtle than that. There are seven hundred meters of shallow-water *Massiccio*, so the continental crust must already have subsided by that much, with the photosynthetic organisms producing sediment fast enough to keep the sea floor shallow. At the end of *Massiccio* time something happened to the sediment producers. Their output fell below the subsidence rate, so the sea floor sank farther and farther below sea level, like the mythical city of Atlantis.

What was that sea floor like? To begin with, light could not penetrate that far through the ocean water, so the bottom was pitch black. In the absence of light, photosynthetic organisms could not live, so there was no abundant basis for a food chain. The only animals that lived down there survived by scavenging the small amounts of dead organic matter that settled down from the sunlit shallow parts of the sea, far above. Even the sediment that was accumulating on the sea floor drifted down from above, and this is the origin of the pelagic limestone.

Mostly these sediment grains were tiny fossils made by single-celled organisms—foraminifera and coccoliths, and the siliceous microorganisms called radiolarians—that floated in the well-lit waters just below the ocean surface. The microfossils sank to the bottom after their makers died. But there was also a component of clay and silt that was carried by ocean currents from the mouths of rivers out into the deep ocean, or blown by winds as dust storms originating in desert areas. Some of the clay and silt was rich in iron that rusted in the oxygen-rich ocean water and gave the sediments of the *Scaglia rossa* their striking pink-and-red colors.

The deposition rate of pelagic limestones is so slow that it could never bring the sea floor back up into the zone where light could penetrate, and

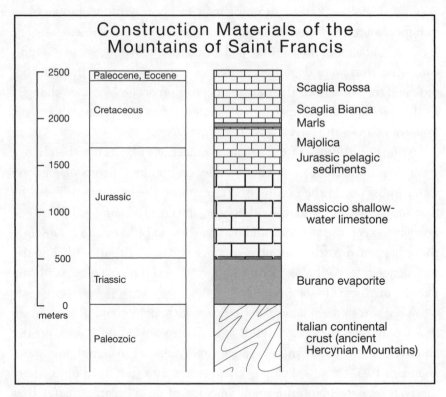

Here are the building blocks from which the Mountains of Saint Francis were constructed. Geologists often use the brick symbol for limestones. Much of the Apennine deformation took place by slip on weak layers, shown in gray, where upper parts of the rock sequence could slide over the lower parts. In chapter 10 we will encounter the additional sedimentary rocks that accumulated during the building of the Apennines.

so the area that is now the Mountains of Saint Francis stayed at pelagic depths through the rest of the Jurassic, all of the Cretaceous, and on into the Paleocene, Eocene, and Oligocene. This produced a great thickness of pelagic sediments. Geologists from all over the world now travel to this part of the Apennines because pelagic sediments, deposited too deep to be scoured and disturbed by waves and currents, carry the best record we have of many aspects of Earth history.

And so a thickness of a mile or two of sedimentary rocks slowly piled up on top of the eroded roots of the Hercynian Mountains during the Mesozoic and the first half of the Cenozoic. Here is a stratigraphic column to summarize the nature of these sedimentary units. There is no place in the Apennines where you could measure this entire section, so it has been pieced together from several different places that expose different parts of the stratigraphy.

The character of this pile of rocks has strongly influenced what the Apennines look like today. Partly this influence has acted through the way different levels deform. Thick-bedded limestones like the *Massiccio* are difficult to fold, but thin-bedded ones like the *Scaglia* flex like a deck of cards slipping over one another.

Some layers are weak and easy to deform, so sliding is concentrated in those layers. We already saw that there is a basal layer of "grease" formed by the Burano evaporites, allowing the rocks above it to detach from the continental crust below.

In addition there is another weak layer within the pile of Apennine rocks, most of which are strong limestones. This second weak layer is the clay-rich interval called the Fucoid marls, deposited during the Middle Cretaceous, and already familiar from the stratigraphic section at Gubbio.[11] This interval, weak because of all the clay in it, allows the Upper Cretaceous and Cenozoic to detach and move across the limestones below it.

The Fucoid marls have played another major role in shaping the Apennines, because the clay they contain makes it easy for them to erode. The limestones above and below are much more resistant to erosion, so the marls have been sculpted into valleys, while the limestones stand up in cliffs and ridges. It is very easy to see this character of the landscape

This view looks northwest along the large anticlinal fold toward Monte Nerone. The Fucoid marls are weak, and erosion has stripped them off the top of the resistant Majolica. *The* Scaglia bianca *and* Scaglia rossa *are therefore only preserved on the flank of the fold, where erosion has not yet removed them.*

as we approach the Burano Valley along the Via Flaminia. The Fucoid marls have been stripped off the top of the anticline, exposing the smooth upward curve of the limestones beneath them. The overlying limestones of the Upper Cretaceous *Scaglia* are preserved only on the flanks of the anticline. Between the two limestone intervals is a cleaned-off surface where the Fucoid marls have been scoured out by erosion. Similar "*Fucoidi* valleys" are recognizable all over the Mountains of Saint Francis, so you can find where you are in the sedimentary sequence with a glance at the landscape.

In the Apennines, once we get above the shallow-water *Massiccio*, we are in the realm of pelagic sedimentation on a fairly deep sea floor. What was going on down there? Mostly it was simply the steady, slow accumulation of sediment drifting down, grain by grain, from above. But looking carefully at the rocks, we sometimes recognize other interesting things that were happening.

For one thing we can infer that there were scavengers on the sea floor, feeding on the gentle shower of dead organic matter from above. They had no hard parts to be left as fossils, but we can see how they disturbed

the sediment as they burrowed through it in search of any bit of nourish-
ment. At the times when the bottom waters were poor in oxygen, no
bottom dwellers could live there and churn up the bottom mud. In these
oxygen-poor waters, the iron-rich minerals did not rust to a red color. As
a result, in the parts of the *Scaglia* that are white or gray instead of red, the
sediment has a very clear and undisturbed layering, reflecting rhythmic
changes in the ocean conditions.

But where the *Scaglia* is red, it is telling us of times with oxygen-rich
waters where bottom dwellers could survive, and the sediment lamina-
tions have been disturbed or even completely destroyed by their burrow-
ing. Sometimes we can even see remarkable casts of their excavations,
showing in detail how they lived, as in the case of the spiral burrows called
Zoophycos.

And finally we can see that the sea floor was not always quiet in our
part of Italy during the long history of pelagic sedimentation. From time
to time submarine landslides broke loose, and the soft mud of the ocean
bottom slid downhill, bending into folds and stretching and tearing apart
like taffy.[12]

These slump folds are con-
fined to thin layers, unlike the
folds that took shape as the
Apennines were constructed,
affecting all the beds in thick
intervals. The folds and swirls
of these slump horizons stand in
striking contrast to the otherwise

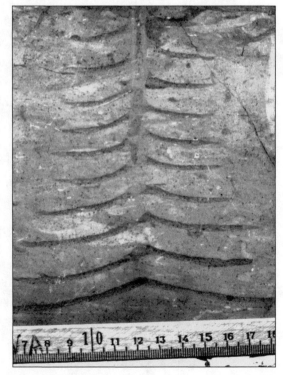

*This Christmas-tree-shaped pattern is
a slice through a spiral burrow called
Zoophycos, made by an animal that lived
down in the soft sediment and stuck its
head out to scavenge dead material that
had fallen onto the deep sea floor. The
scale is in centimeters, and the picture
covers most of one bed of Scaglia rossa,
from top to bottom.*

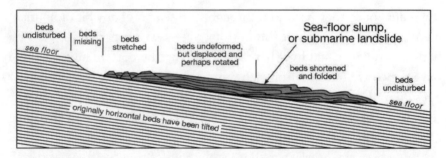

A soft-sediment slump, or submarine landslide, in the sea-floor lime mud that was to become the Scaglia *limestone. In the Apennines we can see all the different parts of slumps like this, though rarely in the same outcrop. The presence of slumps in the Scaglia shows that even in this geologically quiet setting, occasional earthquakes could jar loose masses of sediment on slightly tilted parts of the sea bottom.*

parallel layers of the beds deposited on the quiet sea floor. The slumps tell us that the generally flat sea floor must have had steep slopes in a few times and places, where landslides could break away and move downhill. This shows that the Italian continental crust was not entirely asleep, even in the calm times from the Jurassic to the Eocene. There must have been some slight displacements on nearly dormant faults, tilting the sea floor into slopes suitable for slumping, and perhaps triggering the slumps when earthquakes on those faults shook the sea floor.

The slumps in the pelagic sediments are just minor interruptions to the generally calm conditions, but they hint at the huge convulsions that were to come. As a final preparation for understanding those convulsions, let us look, in a broad view, at Adria, the great block of continental crust that provides the framework for the Alps and the Apennines.

Adria, the Foundation for the Alps and Apennines

Jim Channell, who had worked with us on paleomagnetism at Gubbio, and his Hungarian friend Ferenc Horváth, wanted to understand the setting in which the pelagic sediments like the *Scaglia* were deposited. Jim knew the Italian geology, and Ferenc was familiar with the geology farther north. Piecing together evidence from several countries, they showed that in the Jurassic, Cretaceous, and Early Cenozoic there was a great promontory of African continental crust pointing northward into the Tethyan Ocean that then separated Africa from Europe. They called this African crustal promontory "Adria," and noted that Émile Argand had recognized it long before.[13]

The remarkable concept of Adria first helps us to understand the sedimentary rocks deposited in the Apennines, and then goes on to help explain the mountains of Italy. In order for the deep-water, pelagic limestones of the Apennines to be deposited, the site needed to be protected from any input of clay and sand, for these can accumulate fast and would have swamped the slow buildup of pelagic limestones. Fortunately for those of us who have studied those limestones, the African Promontory stuck way out into the Tethyan Ocean, far from the rivers that then drained Africa, so it stayed free from sand and clay, and mostly limestone was deposited.

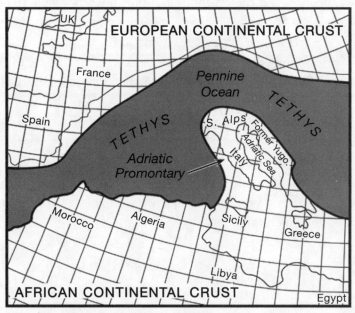

Late Cretaceous (about 85 million years ago)

Jim Channell and Ferenc Horváth made this map to show the configuration of today's Mediterranean region as it would have been in the Late Cretaceous, during deposition of the Scaglia rossa limestone. The gray area is oceanic crust, and the areas marked by latitude and longitude lines are continental crust. The north-pointing spur of African continental crust is "Adria," the Adriatic Promontory.

Bruno D'Argenio, who knows the Apennines very well, joined forces with Jim Channell and Ferenc Horváth and they worked out in detail what the promontory was like, and how it shaped both the sediments and the mountains that were to come.[14] They even found a nearly perfect modern analog to Adria. If we turned the map of Adria around so that north points south, the promontory would look much like the promontory of North American continental crust that contains Florida and the Bahamas and today points south into the Atlantic Ocean and Caribbean Sea. This North American Promontory is far from the sand and clay brought down by the Mississippi River to its delta, so in Florida and the Bahamas, limestone is being deposited today, just as it was on Adria in the Cretaceous.[15]

The Adriatic Promontory is thus the fundamental basis for understanding what Italy was like before it was caught up in the deformation that produced the Alps and the Apennines. Just as limestones like the *Calcare Massiccio* and the *Scaglia rossa* are the small-scale building materials of the Apennines, Adria is their large-scale foundation.

The continental crust of Adria must be down there, under the Mountains of Saint Francis, but it is deeply buried and quite unreachable. Surprisingly, however, there *is* a place not far away where the deep Adriatic crust has been turned up, nearly on end, and eroded off so that we can see it in the field. Let us go and visit that place—the great mountain range of the Alps that closes Italy off from the rest of Europe—to see the continental crust of Adria and to understand why it was brought up where we can see it. After a visit to the Alps, we will be in a better position to understand the Apennines.

9

DISTANT THUNDER
FROM THE ALPS

ONE JUNE DAY in 1985 the train carried us northwestward along the eastern foothills of the Apennines to the city of Bologna, famous for its cuisine and its ancient university—"Bologna, well-fed and well-read," the Italians say. Bologna has long been a major center of geological research. The word "geology" itself was invented there four centuries ago, and the Bolognese geologists still continue that long tradition.[1] There Milly and I met up with my student Dave Bice and with an old friend, Alberto Castellarin, professor of geology at the university and one of the leaders in the study of the Alps. Alberto had agreed to show us some special Alpine geology, to provide a framework for understanding the Apennines.

We drove northward across the flat plain of the Po River, with the Alps gradually emerging from the distant haze. After a family picnic with Alberto's brother near Verona, we started up the great Alpine valley of the Adige River. The new highway leading up the Adige Valley to the Brenner Pass and to Austria was a reminder of the critical role that the Alps and the Alpine passes have played in European history.[2]

The mountain range of the Alps closes Italy off from northern Europe, and has long imposed a major geographic constraint on European his-

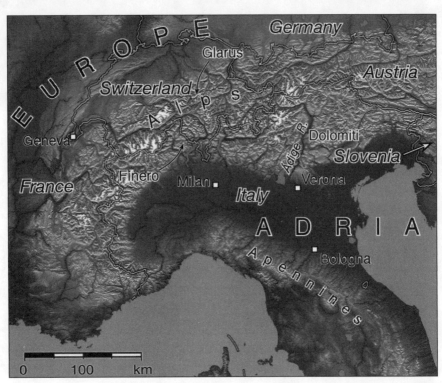

The Alps belong to six major countries—France, Switzerland, Germany, Austria, Slovenia, and Italy. The map shows that these mountains are on a grander scale than the Apennines and act as a barrier, closing Italy off from the rest of Europe. It also shows the locations of the Dolomites, Finero, the Adige River, and Glarus, discussed in this chapter. We shall see that the Alps formed when "Adria," including Italy, was driven northward over Europe.

tory, separating the French and Germanic North from the Italian South. Although small in area compared to the Himalayas or the Andes, these are high, wild, and serious mountains, barricaded by glaciers, threaded by narrow canyons, and threatened by rockfalls and avalanches. Until the first railroads were pushed through in the latter part of the nineteenth century, the Alps were difficult and perilous to cross.

The Alpine Passes

History records the names of the many commanders over the centuries who crossed the Alps, with terrible hardship, leading armies based on one side to cause trouble on the other. Hannibal the Carthaginian crossed into Italy with his elephants to bedevil Rome on its home ground. Julius Caesar crossed northward to begin the Roman conquest of Gaul. As Rome weakened, Alaric the Visigoth crossed the Alps and sacked the eternal city in A.D. 410. Medieval kings like Charlemagne in 800 and the Holy Roman, or German, Emperor Henry VII in 1310 traversed the Alps to be crowned by, or to challenge, the pope in Rome. And the French made the Alpine crossing into Italy in 1796, where the military genius of twenty-seven-year-old Napoleon Bonaparte first became evident.

Many military crossings of the Alps were made in the Middle Ages, some under the appalling conditions of winter, and every one of these traverses of the Alpine barrier was a heroic logistical achievement in itself. In medieval times, in between the occasional military campaigns, there was a constant traffic—sometimes a trickle, sometimes a stream—of merchants, clerics, and pilgrims negotiating the difficult Alpine passes, heading south toward the cities of northern Italy and onward to Rome, and then back to the north.

The Alpine passes played a central role in one of the most dramatic episodes in medieval European history—an episode that ties in to our story of the Apennines, centering around three truly remarkable historical figures. In 1073 the strong, uncompromising papacy of the High Middle Ages was inaugurated with the elevation to the papacy of the fierce Benedictine monk Hildebrand, who took the name Gregory VII. With a plan to reform a degenerate papacy, Gregory was determined to stop a long-established practice called lay investiture, in which the German emperors selected the bishops and abbots within their realm. For Gregory this was an intolerable interference by laymen in the affairs of the church, and he sent a letter to the current emperor, Henry IV, ordering him to cease the practice immediately. Henry, for whom investiture was a traditional right and a crucial source of power, rejected this demand in a critical and insulting reply to Pope Gregory.[3]

Gregory's response was to declare Henry deposed from his throne and to threaten the German bishops and abbots with excommunication if they continued to support him. In the face of this onslaught Henry's power collapsed, and his only possible salvation was to seek papal forgiveness.

This required a truly heroic traverse of the Alps. With the direct passes over the Alps held by supporters of Pope Gregory, Henry and a small party were forced to detour far to the west and to cross the Alpine barrier through the Mont Cenis Pass, in present-day France. It was midwinter of 1076–77, and the snow and ice were appalling. A contemporary account tells of the local guides hired by Henry, of the perilous slipping and sliding on the ice fields, with the men "now crawling on their hands and feet, now supporting themselves on the shoulders of those in front; now and again, as their feet slipped, falling and rolling." It tells of the maiming and death of horses that slid down into canyons, and of the lowering of the queen and her ladies down the mountainside on ox skins.[4] In our day, when the winter Alps are a recreational playground, it is sobering to read what they were like a thousand years ago.

Reaching the plains of northern Italy, Henry hastened to intercept the pope, who was journeying north to reorganize the Germany of the defeated emperor. Their paths met at a castle called Canossa, in the Apennines near Reggio Emilia, the center of power of the third great figure in the drama. Countess Matilda of Canossa was the ruler of Tuscany, when Tuscany was much larger than it is today. Many consider Matilda to have been the most important woman in the history of Italy, and she was a strong supporter of Pope Gregory. Gregory was staying with her at Canossa, waiting, one may imagine, for the summer snowmelt to open the Alpine passes.

In a tense standoff that has captured the dramatic imagination ever since, Henry professed sincere repentance and pleaded for forgiveness, while Gregory, unconvinced, refused the emperor both admittance to the castle of Canossa and absolution of sin. Henry is said to have waited in penitence outside the castle, coatless in the snow, for three days. Finally Matilda persuaded the skeptical Gregory that as pope, he simply could not refuse forgiveness to a truly repentant sinner. Absolved, Henry returned across the Alps to Germany, reorganized his realm so that he would never again be vulnerable to this kind of pressure, and renewed the conflict with the pope, which continued for many years.

Historians have suggested that the German investiture controversy was a critical event in the evolution of Germany, delaying the formation of a united, centralized Germany until the nineteenth century. It has even been suggested that the Gregorian reforms and the investiture contro-versy mark a revolution in worldview equivalent in importance, and very similar in character, to the great modern revolutions of the Reformation, the French Revolution, and the Russian Revolution, in determining the character of Western society.[5] This episode is also a reminder of the central role of the Alpine barrier in European history.[6]

"Mountain Gloom and Mountain Glory"

On that June day, as we traveled farther into Alberto Castellarin's beloved Alps, we saw the mountains as majestic scenery, rich in a beauty that restores the soul. But this was not always so. Before modern transportation, as the story of Henry IV at the Pass of Mont Cenis should remind us, the Alps made travel a harrowing and sometimes fatal undertaking, and mountains did not appear so attractive.[7] Descriptions of mountains three hundred years ago were not flattering—"they were seen as 'Nature's Shames and Ills' and 'Warts, Wens, Blisters, Imposthumes' upon the otherwise fair face of Nature."[8]

Marjorie Hope Nicholson's study of the changing aesthetic view, from "Mountain Gloom" to "Mountain Glory," makes the important point that our present view of the Alps and other great mountain ranges as inspiring and uplifting is the result of the Romantic movement—a fairly recent revolution in taste. Perhaps we could trace the beginning of this aesthetic appreciation of mountains to Albrecht von Haller's 1729 poem *The Alps*, or to Horace-Bénédict de Saussure's 1787 ascent of Mount Blanc, but by the nineteenth century it was in full flower, with romantic painters portraying the Alps and mountains elsewhere in all their seasons and moods.

Nicholson also points out that the emerging science of geology played a critical role in fostering that aesthetic revolution.[9] As a geology major in college I once worried that learning the science of mountains might make them less beautiful in my eyes. That turned out not to be the case for me, and historically the science of geology actually helped to make the mountains beautiful!

Alberto took us into the Dolomite Alps, and we walked up to the pass of San Nicolò, for a view of the great peak of La Marmolada, glorious to our modern eyes. From there we could appreciate the role of geology in the changing of taste that had made La Marmolada seem beautiful. The early geologists lived in a time when Earth history was believed to have lasted only a few thousand years, based on the biblical account of the

The peak of La Marmolada in the Dolomite Alps of Italy—the Alps are beautiful to us today, but were seen as an ugly pile of "warts, wens, blisters" until at least the eighteenth century.

Creation, so they were compelled to see mountains like La Marmolada as the depressing wreckage of some huge and terrifying catastrophe.[10] And at first glance, it does indeed look confused and chaotic. But geologists like Alberto have worked out the detailed story of the mountain's history— a long saga stretching over hundreds of millions of years—and to us it looked like an intricate and lovely sculpture, not a depressing wreck.[11]

Rocks from Great Depths, Exposed to View

The geological story Alberto showed us in the Alps was really remarkable, and it gave us a valuable perspective on our Apennines, three hundred kilometers to the south. He started us with a familiar old friend, the boundary between the Cretaceous and the Tertiary. In the *Scaglia rossa* at Gubbio, the record across that boundary is complete because of continuous sedimentation, but in similar deepwater limestones in the foothills of the Alps at Nago, near Lake Garda, Alberto has found that several million years of history are missing from the record because no sediment was deposited.[12]

As we moved deeper into the Alps, Alberto led us back farther in time, showing us the Alpine Cretaceous and Jurassic rocks, which are not so different from the Cretaceous and Jurassic we knew in the Apen-

nines. But then we entered a new world, for the great peaks of the Dolomite Alps, like La Marmolada, are made of Triassic rocks, which are only
very rarely exposed in the Mountains of Saint Francis. In the spectacular
mountain landscape of the Dolomites we were seeing an ancient Triassic
sea floor exhumed, with modern mountains that long ago had been limestone platforms, rising from the deep sea floor and built up to sea level by
the growth of shallow-water organisms.[13]

Moving down below the Triassic of the Dolomites, we journeyed back
into the Permian and found ourselves in an utterly different world, where
the most remarkable feature is the presence of huge flows of ignimbrites
like those we saw in the Treia Valley near Rome, but much older. These
volcanic rocks rest on top of metamorphic rocks, representing the eroded-
off roots of the still older Hercynian Mountains of Carboniferous time.
Here we had reached the top of the continental crust of the Adriatic
Promontory—far down out of sight in the Apeninnes, but exposed to our
eyes in the Alps.

At this point our trip with Alberto Castellarin came to an end, but if
we had traveled westward through the Italian part of the Alps, we would
have found metamorphic rocks that represent deeper and deeper parts of
the continental crust. Finally, at a place called Finero, northwest of Milan,
near the Alpine lake called Lago Maggiore, we would have come to very
small areas of rock that represent the transition from the lower continental

crust to the top of the mantle
of the Earth, usually far out
of sight, twenty miles down
underneath the continental
crust. It is an amazing sight,
and the Alps provide one of
only a very few places in the

*Alberto Castellarin at Nago, in
the foothills of the Alps, pointing
to a boundary between the white
Cretaceous below and the dark
Eocene above, where no sediment
was deposited on the sea floor for
several million years.*

world where you can walk around on the boundary between continental crust and mantle.[14]

The obvious question immediately arises: Why are rocks that normally lie so deep in the Earth visible at the surface in the Italian Alps? It is clear that the continental crust of Adria and the underlying mantle have been bent upward in a giant flexure, and that erosion has cut down to where we can see even the top of the mantle. But what caused the giant flexure? The answer will emerge as we consider how the Alps came to be.

Alpine Geology and the Mystery of Thrust Faults

The origin of mountains like the Alps has always posed a central problem for geologists, who call this field of study *tectonics*—a word with the same root and same sense as *architecture*. Think how hard a farmer has to work just to move a few blocks of rock out of a field, lift them up, and pile them on top of a stone wall. In mountain ranges something has lifted enormous masses of rock miles into the air and deformed them into complex and intricate patterns. This is not something human beings can do. How does it come about?

The early geologists could not possibly have understood mountains correctly, because the biblical timescale was far too short. By the mid-twentieth century, geologists fully appreciated deep time, but their attempts to understand mountains were shackled by a second misconception—the view that all continents have always stayed in the same places. The German meteorologist Alfred Wegener had argued beginning in 1912 that continents moved around on the globe,[15] but most geologists had rejected that idea by about 1930.[16] Instead of Wegener's continental drift, they accepted the unchanging world map of continental fixism. In retrospect there was never any hope of understanding mountains until continental fixism was overturned and continental drift was appreciated and understood.

The Alps are hallowed ground in the saga of tectonics—the story of how generations of geologists have worked out an understanding of mountain ranges.[17] It has not been easy. Until recently the Alps were difficult of access, and even today many outcrops can be reached only by skilled mountaineers. The Alps stretch across six countries—Italy, France, Switzerland, Slovenia, Germany, and Austria—which have sometimes been at war with one another, and Alpine geologists have thought, written, and argued in Italian, French, and German, although now increasingly in English.

How did the geological understanding of the Alps begin? By the early nineteenth century, geologists knew how to date rocks with fossils, and they knew how to make geological maps showing the distribution across the landscape of various kinds of rocks of various ages. Geologists around the world set out to make those maps, realizing that they would provide the key to understanding how the Earth works, and how it has changed through the long history of deep time. From their mapping the geologists knew the stacking order of different-aged rocks, from older at the bottom to younger at the top.

So imagine the astonishment of the early Swiss geologist Arnold Escher von der Linth in 1840, beginning to map the geology of the Alps, when he first realized that the stacking order of the rocks he was seeing, based on the fossils they contain, was completely wrong! Escher had found a sharp contact in the Swiss canton of Glarus, where Permian rocks rest on top of Eocene rocks. In terms of our modern dates in millions of years, this means that the rocks above the contact are more than six times older than the rocks below it.

At first geologists were completely mystified by the inverted order of rock ages. Even today you can find young-earth creationists who argue that places like Glarus, with their inverted order of ages, demonstrate that geologists do not understand the Earth. The creationists conclude that fossils do not tell us the ages of rocks, that our planet is only a few thousand years old, and that the rocks of mountain ranges like the Alps, with their inverted stacking order, were created all at once, looking just as we see them today. But that view does not hold up if you carefully study the rocks in the Alps and in other mountain ranges.

In the field it is clear that the rocks close to Arnold Escher's anomalous contact are intensely deformed, showing that this is not a surface of deposition in the wrong order but a surface along which older rocks were driven up and over younger rocks that originally lay nearby. Geologists call these sheared surfaces "thrust faults," and their discovery was the key to understanding mountain ranges.[18]

From the time of Arnold Escher up to the present day, thrust faults

Arnold Escher von der Linth's anomalous contact is the prominent nearly horizontal line near the top of the cliffs. Along this surface older Permian rocks lie on top of younger Eocene rocks. This discovery first put geologists on the path to understanding how mountains are built.

Here Milly is pointing to the sharp, horizontal contact discovered by Arnold Escher in 1840 at Glarus, in the Swiss Alps, where Permian rocks unexpectedly rest on top of younger rocks, dating from the Eocene. Note how deformed all the rocks are, with no smooth, horizontal bedding. This is because the Permian was driven up and over the Eocene, damaging the rocks in the process. The sharp contact, marked by the arrows, is a thrust fault, a key feature of mountain belts.

Permian (about 270 million years old)

Eocene (about 40 million years old)

have fascinated geologists, who have studied them in ever greater detail. We have learned to determine the direction of motion of the rocks above the fault and how far they have been displaced. Geologists today can reconstruct the exact paths and emplacement orders of the rocks in a whole pile of thrust sheets, in which the individual thrusts branch and join in a complicated array.[19] Thrust-fault geology is now a highly developed branch of science in which much is understood, although interesting questions remain open. Oil companies use the skills of geologists to understand thrust belts and to find and produce the oil and gas that often lie hidden beneath piles of thrust sheets.[20]

In the Alps it gradually became evident that some of the thrust faults have huge displacements—tens or even hundreds of kilometers. This presented a serious dilemma for the geologists in the middle of the twentieth century. After their mistaken rejection of Wegener's continental drift around 1930, most geologists were firmly convinced that continents have never changed their positions on the map. Only a few—notably the Swiss geologist Émile Argand—saw clearly how well continental drift could explain the Alps.[21]

(It is worth making a point here about scientific evidence. Geologists had long been forced to accept *vertical* movements of the continents, because fossil seashells of animals that lived in the shallow sea near the coastline were found high up in the mountains. The seashells provide an unambiguous marker for ancient sea level, showing that *vertical* movements have unquestionably occurred. But in the absence of a marker for *horizontal* movements, geologists concluded that horizontal movements have not occurred. That was a logical fallacy and a serious mistake, and this experience has led the geologists of today to be very aware that, as we like to say, "Absence of evidence is not evidence of absence.")

Of course, in order to maintain that there were no markers for horizontal motions in the rock record, early-twentieth-century geologists had to ignore the apparent match of the coastlines of Africa and Europe to those of North and South America. Wegener argued that this match *does* provide a marker for horizontal motion. The geologists also had to ignore the evidence for tens or hundreds of kilometers of horizontal displacement along the thrusts of the Alps—evidence that seemed to show that northern Europe and Italy had once been much farther apart than they are now. The Alpine evidence could be ignored because it was vague—it looked like Italy had approached northern Europe, but it was difficult to say, from the thrusts, exactly how far apart they once had been. The thrust displacements provided a minimum distance, but Alpine geology is so complicated that even that could be disputed.

There was even a way of completely rejecting the evidence from the Alpine thrusts for any convergence at all between Italy and northern Europe. This approach was called "gravity tectonics," and the idea was that older rocks have been emplaced over younger ones not by continental convergence but because the old rocks have been uplifted (uplift was allowed, remember!) and have then slid down sideways onto the younger ones.

Livio Trevisan, a major figure in Italian geology at the time, once made gentle fun of gravity sliding in a little cartoon, but it was a serious theory. In the mid-twentieth century, the whole Alpine chain was interpreted as the result of gravity sliding,[22] with little or no continental convergence—a view that had to be rejected as soon as it became clear from other evidence that the continents do indeed move around. And yet, as we shall see at the end of this chapter, gravity tectonics has now made a comeback in a broader vision of how mountains form.

The rejection of the evidence from Alpine thrust faults in the mid-twentieth century is a good example of a phenomenon in science that was pointed out by the astronomers Alan Lightman and Owen Gingerich.[23] They described several other historical cases in which certain observations did not fit a well-established "ruling theory."[24] Lightman and Gingerich called the discrepancies "anomalies." You would think that the presence of serious anomalies in a dominant theory would lead scientists to reject it and look for something better, but well-supported theories are often correct. They develop a life of their own so that it takes a lot—and *should* take a lot—to overturn one. Lightman and Gingerich showed in case after case that anomalies in a strongly held but incorrect theory will be ignored, and not even recognized as anomalies, until the theory is overturned for some other reason. Then the anomalies are "retro-recognized" and seen to provide good evidence for the new theory. Thus it was with the Alpine thrusts—their real significance was not appreciated until plate tectonics replaced continental fixism.

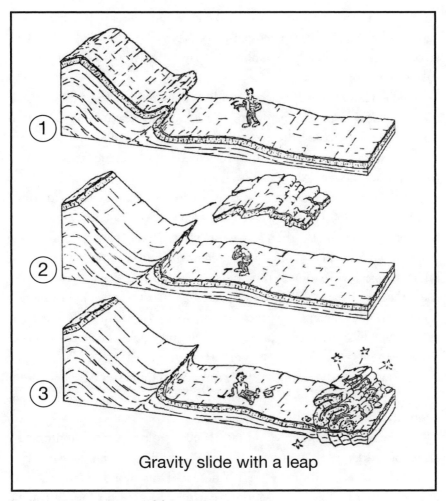

Gravity slide with a leap

Professor Trevisan's "gravity slide" cartoon.

The Triumph of Continental Drift and Plate Tectonics

Continental fixism ruled geological thinking right up to the time I entered Princeton as a new graduate student in 1962. By remarkably good fortune I arrived at just the time and the place where it would all change. Professor Harry Hess accepted me to work on his Caribbean Project, and Milly and I spent our honeymoon in the Guajira Peninsula of Colombia, living with Indians and smugglers and driving a Land Rover around the roadless desert, making a geological map of this northernmost tail of the Andes.[25]

And while I was learning how to make a geological map, the traditional activity that had occupied geologists for a century and a half, my thesis adviser was getting ready to publish the paper that would change geological thinking forever.

Uncle Harry, as we called him, had commanded the Navy assault transport USS *Cape Johnson* during World War II. Leaving his sonar running as he crisscrossed the Pacific, he had discovered strange, sunken, flat-topped seamounts. These must have been old volcanic islands, whose tops had been planed off flat by wave erosion, and then the seamounts had sunk. It was a demonstration that ocean floor can subside with time; and by about 1960 Hess conceived the idea that ocean floor might move *horizontally* as well. By then oceanographers Bruce Heezen and Marie Tharp at Lamont had realized that a great submerged ridge, two or three kilometers below sea level, runs north to south along the centerline of the Atlantic Ocean.[26] Harry Hess speculated that the ridge might mark a source of new ocean crust, with the ocean floor growing in both directions as it moves away from the ridge—like a double conveyor belt—allowing the continents on either side to get farther and farther apart.[27] Geologist Bob Dietz coined the term "sea-floor spreading" for this concept,[28] but for a couple of years most geologists who heard about it considered it just an oddball idea.

However, by the middle and late 1960s, geologists were finding new kinds of evidence that proved sea-floor spreading to be correct. This included the recognition by Fred Vine and Drummond Matthews, and independently by Lawrence Morley, that new ocean crust forming at the axis of the mid-ocean ridge records the direction of the Earth's magnetic field.[29] As a result reversals of the field produce magnetic stripes on the ocean floor, making it possible to reconstruct the positions of ocean crust and the bordering continents back through time. This turned out to be critical for understanding the Alps, as we shall see in a moment. About a decade later Bill Lowrie and I, and Al Fischer's group, were able to date a hundred million years' worth of those reversals through our work on the *Scaglia* at Gubbio.[30]

Another direct confirmation of sea-floor spreading came in the late 1960s, with the first few cruises of the drilling ship *Glomar Challenger*.

Crossing the Mid-Atlantic Ridge, the shipboard geologists drilled a series of holes that showed that the farther from the ridge you are, the older is the sediment that lies directly on ocean crust. The crust is systematically younger closer to the ridge, just as sea-floor spreading had predicted.

Sea-floor spreading was thus confirmed, and it led to a whole set of other discoveries relevant to the Alps. If new ocean floor is being generated at mid-ocean ridges, old ocean floor must be disappearing elsewhere.[31] Geologists soon recognized that the deep ocean trenches, especially around the rim of the Pacific, mark the sites where old crust is subducting—where it is sinking down into the deep-Earth mantle.[32]

Thus an ocean can close up, if its crust is subducted, bringing its bordering continents toward each other. But continents are too thick and buoyant to subduct. So in this new view of continental mobilism—the rejection of continental fixism—it is inevitable that continents will collide.[33] Geologists thus recognized that these collisions would generate great deformation in the leading edges of the two continents. Here was an explanation for mountain ranges that lie between two continents—as the Himalayas lie between Asia and India, and the Alps between Europe and the Adriatic Promontory (Italy).

As geologists in the 1960s and 1970s put together the pieces of this new worldview in an atmosphere of exhilarating excitement, they could see that pretty much all the active deformation of the Earth's surface was concentrated along very narrow bands—at the axes of mid-ocean ridges, at the trenches, and along big transform faults, like the San Andreas, that link up ridges and trenches. Seismic maps showed that almost all earthquakes take place along these bands.[34]

The earthquake maps clearly implied that the broad areas between bands of earthquakes do not deform very much—that they are like rigid caps riding around on the surface of the Earth. Geologists started calling these rigid caps "plates," and they called the bands of earthquakes that separate them "plate boundaries." Most of the dozen or so large plates contain both continental and oceanic crust—like the North American plate, which comprises North America itself as well as the western half of the North Atlantic Ocean. Eventually everyone accepted the name "plate tectonics" for the whole complex of new ideas that were revolutionizing geology.

How Plate Tectonics Solved the Alpine Mystery

As long as continental fixism reigned supreme, the Alps would remain a mystery. Attempts to explain the Alpine thrust faults in terms of vertical uplift and gravity sliding were doomed to frustration and failure. But when plate tectonics came along, it provided the concepts and techniques that made sense of the Alps.

Gradually, through the twentieth century, the basic architecture of the Alps became clear. We can appreciate this architecture if we switch from thinking about the thrust-faults that formed during the deformation, and focus instead on the nature of the rocks that are separated by the thrust-fault surfaces, as the Permian is separated from the underlying Eocene by the Glarus thrust plane.

It is a remarkable pattern, and to visualize it, I find it helpful to think of the Alps as a geological sandwich with three main components—two

This view, looking east toward Austria from the Swiss Alps near Klosters, shows the continental crust of Adria, thrust over the oceanic crust of the Pennine Ocean that once separated Europe from Adria. The European crust is deeply buried here, but it is visible farther west in Switzerland. The strong continental crust of Adria, resistant to erosion, makes the rugged highlands above the thrust, and the weaker crust and sediments of the Pennine Ocean make the gentler slopes lower down. By good fortune the snowline closely corresponded to the thrust when this picture was taken in late September.

slices of bread and a layer of filling. Simplifying the enormous complexity of the Alps, we can recognize that the lower slice of bread is made of the continental crust of Europe, together with the sedimentary cover that was deposited on that crust. The upper slice of bread is the continental crust of Italy—that is, of Adria—the Adriatic Promontory of African continental crust, along with *its* sedimentary cover. The filling in between is the remnants of a smeared-out and intensely deformed oceanic floor we call the Pennine Ocean. This is one part of an ancient ocean geologists call the Tethys, that once separated Italy from Europe, but has completely disappeared as the two continents converged and then collided.[35]

Prior to the Eocene, these three domains—Europe, the Pennine oceanic realm, and Adria—lay adjacent to one another, side by side. Europe was in the north and Italy in the south, separated by the Pennine Ocean. Now they are stacked vertically. The Alps are an impressive example of what the Earth can do. Imagine one continent driven over another continent, smearing out, crushing, and completely eliminating the ocean floor that once separated them![36]

On virtually all the Alpine faults, the upper block has moved northward, and on the scale of the whole mountain range it is clear that the Italian continent has been driven northward over Tethyan ocean floor, which in turn was driven northward over the European continent. This is just what plate tectonics led us to expect. When the ocean crust of one plate is being subducted at a trench, it slides down under the adjacent plate. If a continent belonging to that plate enters the trench, it will be subducted a short distance before its buoyancy brings the subduction to a halt.

So the Alps now made perfect sense. The European continental crust must have been part of the plate that was being subducted, and thus it lies under the Italian continental crust of Adria. All the Pennine oceanic rocks, scraped off the subducted oceanic crust, are caught in between. It was this driving of Italy over Europe that tilted up the crust of Adria, exposing all the normally deep-seated rocks that Alberto Castellarin showed us.

But a major problem remained—how much displacement is represented by the complicated array of Alpine thrust faults? How much Pennine Ocean has been subducted into the mantle, and how far has Adria overridden Europe?

In a remarkable development, plate tectonics also brought a way of solving this problem—a way to determine just how much convergence there has been between Italy and northern Europe. When we make the fit of Europe and Africa against North America, it gives us the former position of Africa relative to Europe, and of course we know their relative positions now. But it is also possible to construct maps showing the positions of the continents around the Atlantic Ocean at several intermediate times, as the Atlantic gradually widened.

This is possible because of the magnetic stripes on the ocean floor. These stripes record the reversals of the Earth's magnetic field—the same reversals we dated with foraminifera at Gubbio. Each reversal is recorded by two symmetrical magnetic stripes, one on each side of the mid-ocean ridge, where the sea-floor spreading takes place. So each pair of stripes can be fitted together, removing all the younger ocean floor between them and giving a map of the positions of the continents at the time that reversal took place. When I arrived at Lamont, two of the geologists there, Walter Pitman and Manik Talwani, were making just such a map, and it showed how the continents of Europe and Africa have moved away from an initial configuration in which both were adjacent to North America.[37]

The next step came when John Dewey teamed up with Walter Pitman, Bill Ryan, and Jean Bonnin, and they realized that the Pitman-Talwani map held a fundamental key to understanding the complex tectonics of the Mediterranean. They recognized that if you consider North America to be the starting point for Europe and Africa and then track those two continents away from North America, you also get the positions of Africa (and its Adriatic Promontory) relative to Europe through time.[38]

This was just what they needed to evaluate the problem of how much convergence between Italy and Europe has taken place on the Alpine thrusts. Beginning with the landmark paper of John Dewey's group, geologists have made continuing efforts to evaluate this convergence. Here is one by Jim Channell and Ferenc Horváth, which I have chosen because it is very clear, and focuses on the Adriatic Promontory.

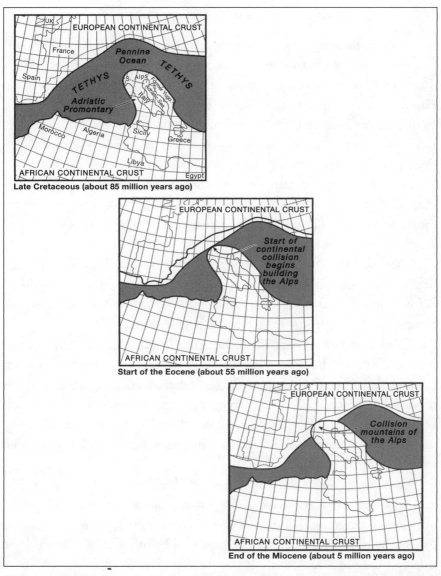

The approach of Adria, the Adriatic Promontory of African continental crust, toward the European continent, based on knowing the motions of Europe and Africa away from North America. When the two continents collided, Adria overrode Europe, and the Alps were the result. Geologists call this kind of ultrasimplified diagram a "cartoon." It focuses on the roll of Adria in Alpine deformation, omitting all the formidable complexities of the real geology.

The Broadest View of Mountain Tectonics

In our examination of the Alps we began by looking at the whole mountain belt as a great topographic barrier between Italy and the rest of Europe. Then we zoomed in on two details—the deep rocks exposed in the Italian Alps, and the thrust fault at Glarus. Finally we expanded our view still farther and recognized the whole mountain range of the Alps as the site of a collision between two great blocks of continental crust—Europe in the north, Italy and the rest of the Adriatic Promontory in the south.

Let us finish by tying this plate-tectonic view of the Alps to the topographic view we started with. How do colliding continents generate a mountain barrier? The answer is new and surprising, and offers a profound insight into how our planet evolves.[39]

If we view the Alps in the broadest perspective of distance and time, details like thrust faults fade into a blur of smooth, almost continuous deformation.[40] In this view the Alps behave like a pile of sand on the beach, which can be squeezed, molded, and sculpted.

Think about what happens as Europe and Italy collide. Their continental crust is too buoyant to sink into the deep Earth, so as the collision continues, deformed crust must pile up higher and higher in the collision zone. When we realize that there have been many hundreds of kilometers of shortening in the collision, it is clear that the pile of deformed crust would get very high—far higher than the Alps are. What has happened to all that piled-up crust?

There are two ways that the pile is kept from growing ever higher. The first is by erosion, which attacks the growing mountains and carries them away, grain by grain. The higher the mountains, the more vigorous the erosion by rivers and glaciers. The second is by spreading out toward the sides, on the north and south. If you try to push sand on the beach into a pile that is too high, the pile will collapse outward. Gravity drives this collapse, and so we see the rebirth of gravity tectonics in a new form. Gravity carries the rising mountains away, thrust sheet by thrust sheet.

Geologists today are beginning to see the topographic barrier of the Alps as held in a delicate, constantly changing balance between the rise of continental rocks caught in the collision zone, their removal by erosion, and their outward collapse, driven by gravity. The balance is very sensitive, and even changes in rainfall can affect the evolution of a mountain range. There is evidence that wetter periods have allowed erosion to keep the Alps from getting too high, but when rainfall and snowfall decline, erosion is no longer effective, and great gravity-driven thrust sheets are needed to carry the rocks of the Alps down and outward to the north and the south.[41] This very dynamic view of landscapes is now being applied to mountain ranges all over the globe.[42]

In the Eocene the great collision that would produce the Alps was just beginning. A few hundred kilometers to the south, in the part of the Adriatic Promontory that was to become the Mountains of Saint Francis, the rock record in the *Scaglia* continues undisturbed through the Eocene and contains no hint of the Alpine collision. To Apennine ears, the wrenching deformation of the Alps was only distant thunder. But the long quiet times were coming to a close. In just a few more million years our quiet Italian sea floor would be overwhelmed by debris from an approaching wave of tectonic deformation. Then the Earth storm itself would hit, driven by hidden processes below the crust, deforming the accumulated sediments of the tranquil sea floor into today's Apennine Mountains.

So now let us return to the Apennines and see how they were built.

10

THE APPROACH OF DESTINY

AFTER MILLY AND I found the Bottaccione road closed by snow in December 1970, we learned of a new pass over the mountains east of Gubbio, a road up the nearby Contessa Valley. As we drove up the Contessa road, we were again—as in the Bottaccione an hour earlier—passing outcrops of rocks that would become familiar old friends in years to come.

From time to time in those years we would wonder who she was—who was the countess for which the Contessa Valley was named? At some point I heard it suggested that the valley was named for the very same Countess Matilda of Canossa who hosted in her castle the epochal confrontation between Pope Gregory VII and Holy Roman Emperor Henry IV in the winter snows of 1077. Could that be the case? Historical maps of Matilda's time show her realm of Tuscany ending at (or very close to) Gubbio, so it is possible that the Contessa Valley marked her border. But my friends who have searched can find no documentary evidence linkng Countess Matilda to the Contessa Valley, so the connection may be no more than an appealing speculation. Just as geologists need evidence from the rocks to reconstruct Earth history, so historians require evidence from written documents.[1]

Clay and Sand—the Heralds of Change

The rocks in those two snowy canyons of 1970 turned out to complement each other perfectly in telling the historical story of this part of the Apennines. The Bottaccione Gorge has wonderful outcrops of our old friend, the *Scaglia rossa* limestone, so it records the history of the Late Cretaceous and the first part of the Paleogene. But if we want to understand the younger beds that lie on top of the *Scaglia*, the Contessa Valley has much better exposures.

Let me first show you a simplified stratigraphic column of the rock layers in the Contessa Valley, to point out a set of four intriguing mys-

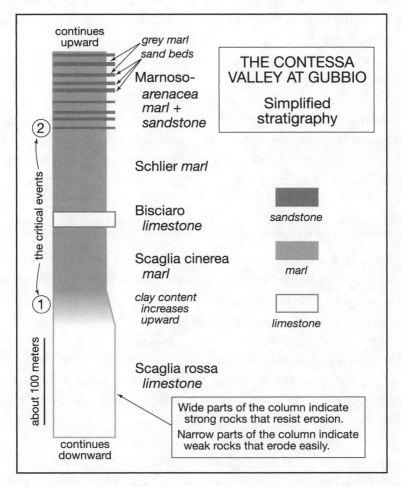

continues
upward

grey marl
sand beds

Marnoso-
arenacea
marl +
sandstone

THE CONTESSA
VALLEY AT GUBBIO

Simplified
stratigraphy

Schlier *marl*

the critical events

Bisciaro
limestone

sandstone

Scaglia cinerea
marl

marl

clay content
increases
upward

limestone

about 100 meters

Scaglia rossa
limestone

Wide parts of the column indicate
strong rocks that resist erosion.
Narrow parts of the column indicate
weak rocks that erode easily.

continues
downward

A simplified stratigraphic section showing the rocks exposed in the Contessa Valley near Gubbio. The two key things to notice are (1) the increase of clay that changes the strong Scaglia limestone into the weaker marl above it, and (2) the appearance of the first sandstone bed that inaugurates the Marnoso-arenacea sandstone. These events mark the approach of the Apennine Earth storm toward Gubbio.

teries. By the end of this chapter, after we see how geologists solved those four mysteries, we will understand the whole story these rocks have to tell. It is a tale of the approach of destiny, as the great Apennine Earth storm slowly approached the place where Gubbio now stands.

In broadest outline the Contessa rocks record two critical changes on the ancient sea floor at Gubbio. In the first change the *Scaglia rossa*—last of the nearly pure limestones deposited here through all of the Jurassic and the Cretaceous—gradually became diluted with clay. More and more clay reached the site of Gubbio until the sediment became a marl—a mixture of limestone and clay.

The second change brought something completely new to the Mountains of Saint Francis—the first sandstone bed!

Let me explain why an apparently innocuous sentence like that needs an exclamation point. This is *literally* the first sandstone bed. If we look at that first bed carefully, we can see that it is full of grains of quartz and feldspar—mineral grains derived from the continental crust—from granites and metamorphic rocks. In all the many hundreds of meters of sedimentary rocks downward to the oldest levels exposed anywhere in the Mountains of Saint Francis, there is nowhere, to my knowledge, a single bed of quartz-feldspar sandstone.[2]

Above that first thin sandstone bed, we can see more and more beds of sandstone—many dozens of them, some thin, some thick. Each sandstone bed is separated from the next one by an interval of gray marl that is just like the *Schlier* marl that was deposited before the first sandstone bed. The alternating layers of sandstone and marl make up a thick formation that is very prominent in the Mountains of Saint Francis. In one of the few prosaic formation names in Italian, although a very descriptive one, this formation is known as the *Marnoso-arenacea*—*marnoso* meaning "marly" and *arenacea* meaning "sandy."

To summarize what we see in the Contessa outcrops, tiny shells of floating foraminifera and algae had long settled to the sea floor, together with a small fraction of clay, to make the pelagic limestone at Gubbio. In the first major change the deposits of tiny shells gradually became diluted with more and more clay. Thus the pelagic limestone slowly changed upward into a marl. Then, in the second change, at a certain time, as marl continued to accumulate, a first sandstone bed appeared. And more and

Here, in the Contessa Valley near Gubbio, is the first sandstone bed in 200 million years. Then more and more sandstone beds arrived, making the Marnoso-arenacea *formation, with resistant sand beds protruding, and weak marl beds eroded back.*

more sandstone beds were deposited, at intervals, punctuating the slow accumulation of marl.

What does all this mean? Let us explore this mystery in terms of four questions—each of them a puzzle that geologists have had to solve in order to understand what was going on at this critical juncture in the history of the Mountain of Saint Francis. First, we need to know the age of the *Scaglia cinerea* marl that follows the *Scaglia* limestone, and also the age of the sandstone beds of the *Marnoso-arenacea*. Second, we need to ask why there is no sand in the Apennine rocks lower down, and when the sand did arrive, where it was coming from. Third, we will ask how these beds of sandstone were deposited, a question that was answered only by a revolutionary discovery. Finally, we need to understand why the marl and sand appeared at this particular time at Gubbio, and how they are related to the folds of the Apennine Mountains. Let's consider each question in turn.

How Guido Bonarelli Got the Age of the Sandstones Wrong

Our first question, about the age of the marl and the sandstone, proved to be a trap for Guido Bonarelli, the early geologist of Gubbio. In 1897, before beginning his exploration of wild and remote parts of the globe, young Bonarelli took up a position teaching geology at the University of Perugia, and although he only stayed four years, in that time he made a geological map of a large area surrounding Perugia. On foot and, I presume, on horseback, Guido Bonarelli systematically examined the rock outcrops of this beautiful part of Italy, a land of rolling hills and small farms, with occasional modest mountains. Bonarelli's map was not published at the time, but it was of such high quality that it was discovered and finally printed posthumously sixty-six years later, for even then it was the best map available for the mountains of Perugia.[3]

In the days when Bonarelli was making his superb geologic map, geologists used fossils to determine the ages of rocks—big seashell fossils you can hold in your hand and study with the naked eye. Bonarelli was skilled at dating rocks with fossils, but he had a real problem with the *Scaglia* and with the gray marls and the *Marnoso-arenacea* sandstones—because they just don't contain *any* of those big, visible fossils. So how could he tell their ages?

In chapter 6 we saw that the problem was solved when geologists learned to use microfossils, which are abundant in these rocks. Let us return to the question, for there is a remarkable twist to the story. From his mapping near Perugia, Bonarelli knew that there are other sandstones, which geologists today call the *Cervarola* sandstone, that rest on top of the *Marnoso-arenacea*. In the *Cervarola* sandstones, which are not found at Gubbio, Bonarelli collected fossils of Oligocene age. The obvious inference was that the *Marnoso-arenacea* is older than Oligocene, and thus Bonarelli showed it on his map as being Eocene in age. Carrying this inference downward, he concluded that the *Scaglia* limestone was all of Cretaceous age. It was eminently reasonable but entirely wrong!

In later years, when geologists like Otto Renz, following in the footsteps of Ambrogio Soldani, learned to date rocks with forams, they discovered Bonarelli's mistake.[4] The forams showed that the *Marnoso-arenacea* is Miocene in age, not Eocene! It is much younger than Bonarelli thought. The gray marls of the Schlier are early Miocene in age, and the *Scaglia cinerea* is Oligocene. The upper part of the *Scaglia* limestone is also much younger than he thought. Guido Bonarelli, to his eternal credit, never disparaged these younger, revised ages or tried to defend his earlier work. In the finest tradition of science, Bonarelli fully accepted the new information and used it as the basis for the geological work of the last part of his life.[5]

But if Bonarelli's inferred age for the *Marnoso-arenacea* was wrong, then his reason for the inference must also have been wrong. How could the older Oligocene sandstones of the *Cervarola*, dated with big fossils, lie *on top* of younger Miocene *Marnoso-arenacea*? The answer was clearly that these two sandstone formations are not in their original stratigraphic order. It was exactly the same situation Arnold Escher von der Linth had discovered in the Alps sixty years earlier. The older *Cervarola* sandstones have been driven up over the younger *Marnoso-arenacea* sandstones along a thrust fault. And the thrust faults of the Apennines are absolutely critical, as we shall see in the next chapter. I think Guido Bonarelli would have figured this out if he had had beautiful, clean Alpine exposures to study, but in the gentle farming hills around Perugia, all covered with soil, the thrust faults lay hidden and unrecognized until the study of forams showed that the thrusts just had to be there.

The point of our whole investigation is to understand how the Mountains of Saint Francis came to be. That understanding depends critically on the detailed dating of the rocks. The emerging picture of a great Earth storm of thrust faults sweeping across the Italian Peninsula comes from knowing very exactly the ages of the rocks that witnessed and recorded the passage of the storm. If all we had was Bonarelli's early work, we would not know of the thrusts that brought the Apennines into being. Guido Bonarelli and the geologists of his generation laid the foundation of our knowledge, but only the corrected dates that came from microfossils made possible the detailed understanding we have today.[6]

Why Was There No Sand in Adria?

To address this second question, let us return to the Contessa Valley and savor the mystery that the rocks present to our eyes. Perhaps it is not the appearance of the first sandstone bed that is so remarkable, but rather the absence of sandstone throughout all the layers lower down.

This absence of sandstone is very unusual, because sand is extremely common in the geological record. Rivers erode sand from mountains and bring it down to the sea, and coastal currents carry it along the shore. Sand is found almost everywhere.[7] But about two hundred million years passed in this part of Italy with no sand deposited at all. What was keeping the sand away from Italy?

The reason is now quite clear, and it comes back again to Adria—that fundamental component in the paleogeography—the ancient configuration of lands and seas—that controlled so much of the geologic evolution of Italy. For all that time Italy was an isolated region of shoals, low islands, and deepwater areas on the northward-pointing Adriatic Promontory of Africa, and no rivers could reach this Italian region and bring sand to it. Like modern, southward-pointing Florida and the Bahamas, Adria was simply unreachable by rivers carrying sand.

The long sandstone-free phase ended abruptly at Gubbio in the middle Miocene. The first sandstone bed is thin, and it is not very impressive, but its significance is huge. Some truly major change must have taken place, and the isolated promontory was no longer off limits to sand.

In the light of the previous chapter we can understand what the change was. The collision of the Adriatic Promontory with Europe in the Eocene pushed up the Alps at the northwest end of the Promontory (see the figure on page 165 in the previous chapter). In those newly risen Alps, rivers could erode down through the sedimentary rocks that cover the continental crust of the promontory. Beneath the sedimentary rocks the rivers cut into the granites of the continental crust, and for the first time released floods of quartz and feldspar sand into this formerly isolated area.

So the sand of the *Marnoso-arenacea* comes from the Alps. But the Alps are three hundred miles away. How did the sand travel so far? That question will lead us to one of the great geological discoveries of the twentieth century.

Graded Beds and the Turbidite Revolution

The sandstone beds of the *Marnoso-arenacea* are surprisingly interesting, and for many years they were one prominent example in a really difficult puzzle that mystified geologists all around the world. The puzzle was to explain how sand could get out into deep water. Let us relish both the mystery and its solution, for they involve one of my favorite stories about geologists.

Here was the problem: The *Scaglia* limestone is clearly a deepwater marine sediment, full of planktic foraminifera—ones that lived in the surface waters of the open ocean and settled to the sea floor far below when they died. The sandstone beds of the *Marnoso-arenacea* represent a completely different situation. The *Marnoso-arenacea* sand grains must have been eroded from a mountain range—almost surely the Alps.[8] But how could those sand grains get into the marine realm of the *Scaglia*, far from the Alps or any other mountain range?

Rivers carry sand, but they stop at the coast. Waves move the sand grains around on a beach, but there are no waves in the deep sea. It was a

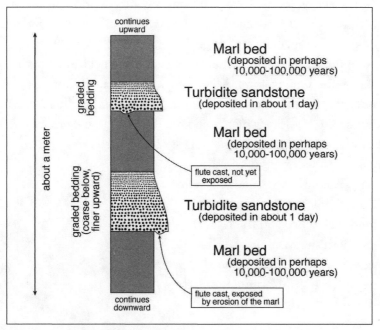

This diagram shows the graded beds of sandstone in the Marnoso-arenacea *at Gubbio, and throughout the Mountains of Saint Francis. The entire* Marnoso-arenacea *contains hundreds of sandstone beds like these. The sandstone beds stick out farther, because they are more resistant to erosion than the marl layers. We will soon see why geologists conclude that each sandstone bed represents perhaps a single day, while the marl beds took many thousands of years to deposit.*

major conundrum for sedimentary geologists. Massive amounts of sand were somehow getting out onto the sea floor, but no one could imagine any mechanism that could carry the sand down into deep water.

Deep-sea sand beds like those of the *Marnoso-arenacea* had been recognized in many places around the world by the mid-twentieth century, and they usually showed a feature that came to be known as graded bedding—in which the sand grains at the base of the bed are coarse, and the sand becomes progressively finer toward the top of the bed.

Each bed is graded, and this is so consistent that in an intensely deformed mountain belt, where some of the beds are upside down, geologists can use grading to tell whether a sand bed is upright or overturned. Grading is pretty much restricted to these deep-water sand beds. Dunes in the desert do not show grading and neither do river and beach sands. What does the grading mean? It would take a minirevolution in geology to answer this question.

The revolutionary breakthrough came just after World War II, when two scientists, completely independently, figured out how graded sand beds are deposited. In Florence the Italian geologist Carlo Migliorini had been carefully examining graded beds in the Apennines in the field and thinking about how the grading might form. Migliorini was particularly interested in the *Marnoso-arenacea* and in a somewhat similar formation farther west, in Tuscany, called the *Macigno*, which is about to become an important part of our story. A *macigno* in Italian is a big, robust block of stone, and the formation gets its name because its resistant sandstone beds were useful in constructing the Renaissance buildings of Florence and the rest of Tuscany. Both the *Macigno* and the *Marnoso-arenacea* are full of graded beds, and Migliorini wanted to understand their origin.

Carlo Migliorini (1891–1953), the Florentine geologist who discovered, through his work in the Apennines, that graded beds are "turbidites," deposited by underwater sediment flows.

Carlo Migliorini was struck by the continuity of the graded sandstone beds, each of which has an almost constant thickness when followed for miles and miles. He knew that rivers and winds, and sea currents near the beach, always deposit sand bodies that vary in thickness over short distances. Think of a river—it has a thick fill of sand in its channel and almost no sand on either side. Geologists also knew that coarse sediment grains are only carried by fast currents, while fine grains can settle out only when the current is slow. How, then, to account for graded beds—coarse at the base and fine at the top?

This and other observations in the field led Migliorini to the idea that the sediment of each graded bed had been brought down into the deep ocean as a cloud of dirty, "turbid" water. He inferred that the sand of the *Macigno* and the *Marnoso-arenacea* had flowed down the gently sloping sea floor into deep water as dense mixtures of sediment and water, with each bed representing one flow. When the flow reached the flat sea bottom and slowed down, the sand settled out, coarsest and heaviest grains first, to form a graded bed. Between the times of sediment flows, fine clay would settle out, grain by grain, to make the clay intervals that separate the sand beds. Migliorini got it exactly right, just from looking very closely at rocks in the Apennines, and thinking carefully about them.

The other discoverer was Philip Kuenen, a Dutch geologist who was doing experiments with wet sediment in tanks of water. Because of the importance of water to a country much of which is below sea level, Dutch geologists have long excelled in this kind of work. Kuenen found that if he stirred up a mess of wet sand and clay, and released it into a tilted tank of water, the mixture would flow down the slope of the tank. The turbid mixture of water and sediment is denser than the pure water in the tank, so it would flow down to the deepest part of the tank. When he examined the deposits of sand from his experiments, he found they were graded.

Kuenen realized that this kind of turbid, dirty sediment flow could carry sand into deep water and could explain the graded sand beds in deepwater settings. He proposed the name "turbidity flow" for the dense mixture of sediment and water, and "turbidite" as a name for the graded sand beds, implying this origin.

Carlo Migliorini and Philip Kuenen happened to meet at a scientific

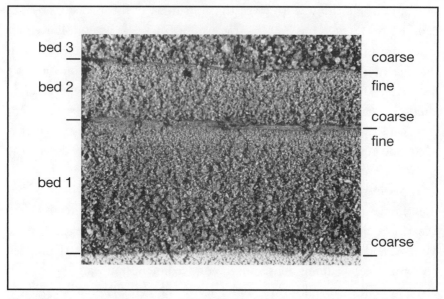

These beds deposited by sediment flows in Philip Kuenen's laboratory are graded, with coarser grains at the bottom and finer above. The grading was just what Carlo Migliorini was seeing in the mountains of the Apennines. When they reported this together, it triggered a revolution in geological thinking.

conference shortly after World War II, as scientists were reestablishing contacts between countries that recently had been mortal enemies. They fell into conversation and discovered that they had independently reached the same solution to the graded-bed mystery.[9] In such a situation a common reaction is for each scientist to rush to publish first. But Kuenen and Migliorini chose a more civilized path—they decided to work together, and together they published a paper that immediately became a classic, triggering a revolution in sedimentary geology. The title makes it very clear: "Turbidity currents as a cause of graded bedding." Their paper, published in the *Journal of Geology* in March of 1950, begins with a nice statement of how the paper came about, as well as a tribute to two other geologists who had had the same idea ten years earlier but whose work had not been noticed, perhaps in the confusion of the war. Kuenen and Migliorini wrote:

> The present authors arrived independently at the conclusion that the
> most important types of graded bedding appear to have been produced

by the action of turbidity currents of high density on the sea floor. But, whereas one of us (Migliorini) was engaged in the investigation of graded rocks encountered in the field, the other (Kuenen) had studied artificial turbidity currents of high density in the laboratory. When they found how closely their conclusions tallied, although arrived at from opposite starting points, they decided to present their results jointly to a wider public. They have since discovered that Bramlette and Bradley (1940) had already indicated the same process as explanation of some graded beds encountered in recent deep-sea deposits.[10]

Graded beds were explained at last! They are turbidites. Kuenen and Migliorini's paper was electrifying, and it set off the turbidite revolution. Sedimentary geologists headed for the mountains to look carefully at turbidites, and one of the best places to look was Migliorini's Apennines.

The new generation of geologists who became interested in turbidites after 1950 learned a great deal from features called "sole marks," common on the bottom surfaces—the soles—of graded, deep-sea sandstone beds everywhere. They show up particularly well in the *Marnoso-arenacea* because the strong sandstone beds, resistant to erosion, contrast with the soft, weak marl beds that easily get removed as water from storms and streams washes over the outcrop. In an exposure of *Marnoso-arenacea* you typically see the resistant sandstone beds protruding, while the shale beds have been eroded back into a recess.

If you look on the exposed bottoms of the sandstone beds, you often see the most striking patterns. The sandstone beds do not have smooth, flat bases. They contain all sorts of bumps and raised lines and protruding, teardrop-shaped ornaments, like abstract bas-relief sculptures. These sole marks come in a wide variety of intricate and sometimes beautiful shapes. Geologists—or anyone else who happens to notice them—can quickly become fascinated and spend hours finding new patterns and trying to explain them.

When you get interested in sole marks you quickly realize that they formed as *casts*. Just before the deposition of the sand bed, something must have cut a variety of molds—grooves and depressions—into the top surface of the underlying marl, which was still a soft mud. Then the sand of the turbidity flow filled the molds and the sand became naturally cemented into hard sandstone. In an outcrop the weak marl of the mold has been washed away, leaving the hard sandstone casts exposed to view. It is exactly analogous to the way a sculptor makes a mold, fills it with molten bronze, and then removes the mold to release the bronze cast, which becomes the final sculpture.

Of the enormous variety of sole marks, the ones called "flute casts" have provided a treasure trove of information. Flute casts look like tongues or teardrops on the base of the sandstone bed, with a sharply defined end in one direction but fading out in the other. Remember that these protrusions are casts made of sand that filled a mold cut down into the underlying marl.

Shortly after the landmark paper of 1950, geologists recognized that each flute cast marks the spot where a swirling, turbulent eddy in the moving turbidity current touched down and scoured out a pit in the soft sea floor—a pit that fades out downstream because the eddy weakens as it scours. Flute casts made it possible to determine the direction the turbidity flow was moving, and this in turn reflected the slope of the sea floor down which the current was flowing. It is a remarkable thing that geologists can reconstruct the motion direction—the paleocurrent—of a flow that lasted

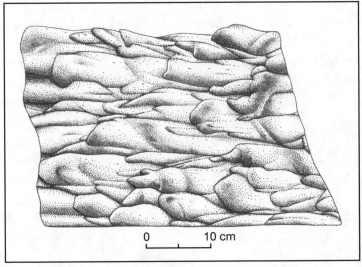

0   10 cm

Flute casts from the Marnoso-arenacea, *from a drawing in the PhD thesis of Ernst ten Haaf. We are looking at the bottom of a turbidite bed turned on end. The flute casts show that the turbidity current moved from left to right.*

Franco Ricci Lucchi describing his Contessa key bed, along the Contessa road near Gubbio. The Contessa key bed—the massive bed behind Franco—is a single colossal turbidite sandstone in the Marnoso-arenacea *that provides a marker that can be traced for hundreds of kilometers.*

maybe a few hours and took place millions of years ago! The turbidity-current revolution gathered steam as geologists realized they could map and analyze the shape and the filling history of ancient deepwater basins using the paleocurrent information from flute casts.[11]

In the first decade of the turbidity-current revolution, a young geologist from the land of Kuenen came to the land of Migliorini to study the Apennine graded beds in light of the new ideas about turbidites. In 1959 Ernst ten Haaf finished his PhD work, showing from flute casts that in both the *Macigno* and the *Marnoso-arenacea*, most turbidity flows had traveled toward the southeast. They were coming from the northwest, so ten Haaf speculated that the sand might originally have been eroded from the Alps, a view that has been fully confirmed. The currents were flowing parallel to the ridges of the modern Apennine Mountains.[12] But how could that be, when the mountains did not yet exist? It was an intriguing mystery that turned out to be a critical clue to the origin of the Apennines.

In the 1960s two young Italian geologists, Emiliano Mutti and Franco Ricci Lucchi, followed up on ten Haaf's initial reconnaissance and began detailed studies of the Apennine turbidites.[13] Over the years their results and those of their students and other geologists who joined in have made the Apennines into one of the most thoroughly understood turbidite provinces in the world, a place where many breakthrough concepts have emerged.[14] Franco has included many pictures of turbidite features in his wonderful photographic atlas, *Sedimentographica*,[15] which is aesthetically appealing as well as scientifically important.

The Beginning of the Apennines

The detailed understanding of the Apennine turbidites that the Italian geologists have built up provides a fundamental key to the puzzle of how these mountains came to be. Around Perugia and Gubbio there are two turbidite units. The *Cervarola* sandstone was deposited earlier and farther to the west. The *Marnoso-arenacea* is just slightly younger than the *Cervarola*, it was deposited slightly to the east, and almost immediately the *Cervarola* was thrust up and over the *Marnoso-arenacea*, driven from west to east.

There must be something significant about the deposition of two different turbidite units, so nearly the same age, with the older one driven eastward over the younger one. In fact this turns out to be a theme of the Apennines, with our Gubbio case being just one example. The Italian geologists have recognized that there are four different turbidite sandstones in the Oligocene and the Miocene, filling four separate sea-floor depressions, in a pattern that gradually migrated from west to east across Italy.

From the *Macigno* sandstones in the Oligocene, through the *Cervarola* and the *Marnoso-arenacea*, to the *Laga* sandstones in the Late Miocene, turbidity flows have carried sand from the eroding Alps southeastward along the trend of the modern peninsula. And as the locus of sand deposition has swept eastward, it has been followed by a locus of thrusting. Soon after turbidite deposition came to an end in each of the troughs, the turbidite sandstones were driven eastward, up and over the sandstones that were being deposited in the next-younger trough. It is clear that the turbidite deposition and the thrust faults are tied together in a fundamental way, in a linkage that is at the heart of the growth of the Apennine Mountains, as we shall see in the next chapter.

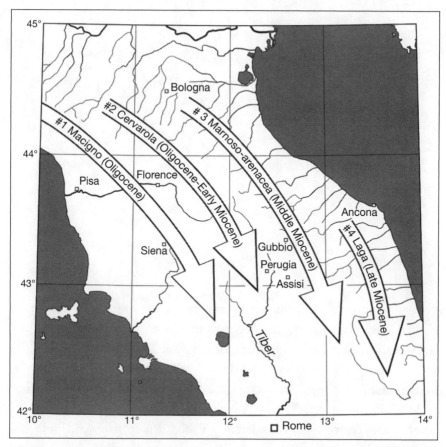

This map shows the pattern of turbidite deposition in what are now the northern Apennines, although the present geography, shown by the rivers and coastline, did not exist at the time. Turbidity currents carried sand from the Alps toward the southeast, but the sea-floor trough in which the sand was deposited gradually migrated toward the east. This pattern is a fundamental key to the origin of the Apennines.

11

Paroxysm in the Apennines

After our brief glimpse of the mystery canyon of Furlo in the winter of 1970, Milly and I spent the night nearby, in the little city of Urbino. The next day, after exploring the artistic treasures of Urbino, home of the Renaissance painter Raphael, we drove through snow-covered foothills down to the Adriatic coast.

In future years I would come to understand that a coherent story tied together the entire winter landscape we had driven through. From the pass at Gubbio to the high ridge cut through by the Burano Gorge, to the cliffs of Furlo, to the foothills of Urbino, and on to the shore of the Adriatic Sea—our track had crossed much of the Apennine Mountain range. As we were finding our way through a winter storm, we were also traversing a landscape that records the passage of a great Earth storm that has left this mountain range as its monument.

Northward along the Adriatic Coast, with the weather slightly improved, we came at last to Ravenna, the goal of our journey. This little city is famous for the Byzantine mosaics that decorate its churches. Protected by a maze of coastal marshes, Ravenna once served as the capital of Italy, when Rome itself lay open to attack. That was during the Gothic War of the sixth century A.D., which for decades ravaged and ruined what was left of Roman Italy, as the Byzantines under the Eastern Roman emperor Justinian tried to recover the land from the

Snowy foothills near Urbino. These hills are an integral part of the Apennines, telling a critical part of the geological story.

Germanic Ostrogoths. The Via Flaminia, which Milly and I had just followed, linked Rome and Ravenna and became the major strategic focus of the Gothic War.

Years later, when my students and I were studying the geology of the Furlo Gorge, we would pass through a little Roman tunnel cut through the cliffs. Long ago the tunnel had been fortified by the Goths, who thus controlled this part of the Via Flaminia. Finally the Byzantine troops climbed up high above the tunnel and rolled down boulders, until the Goths surrendered in terror. The Byzantines had employed a geological solution to a military problem, and we geologists greatly admired it![1]

Coming back from Ravenna, we found that the winter storm had returned with renewed vigor. We had to put chains on the tires before the police would let us cross the Apennine pass at Montecoronaro. There was no hope of visiting a place called Monghidoro, near the Apennine watershed between Bologna and Florence. But I could not help thinking about it, for Monghidoro was the first Italian geological mystery I had ever encountered. I can still remember my sense of wonder when I heard the story.

Surfing the Waves of an Earth Storm

I learned about Monghidoro as a graduate student at Princeton, taking a class from Professor John Maxwell. In the late 1940s, as scientific communication was reopening between nations recently at war, John Maxwell had read about some remarkable geological results from Italy. I still remember him telling our class, with a trace of awe in his voice, about the discoveries of Giovanni Merla at Florence. Maxwell was so intrigued that he just had to go and see for himself, and so he had met Merla and they had become friends and colleagues.

John Maxwell built up the drama as he told us about Merla's discovery. On top of the continental crust of Italy, he told us, layers of sedimentary rocks had accumulated for two hundred million years, from the Triassic to the Early Tertiary. These are the rocks that later became so familiar to me—the *Calcare Massiccio*, the *Scaglia*, the *Marnoso-arenacea*, and all the rest of the sequence exposed at Gubbio. But lying on top of them—here was the wild part—is another suite of rocks that are completely different and have nothing to do with the Italian continental crust.[2]

These foreign rocks, of many different ages, are called the Ligurides, and they were deposited in an ocean to the west of Italy—an ocean that no longer exists. Some Liguride rocks are pieces of ocean crust, and some are the sediments, including lots of turbidites, that accumulated in that Liguride Ocean.[3] These oceanic rocks were driven out of that ocean when it was squeezed shut, and they have slid almost the entire way across the Italian Peninsula! This is not the way geologists usually picture the subduction that closes an ocean. In the Alps the rocks of the closed-up

Giovanni Merla (1906–84), the Florentine geologist who pioneered our understanding of the Apennines.

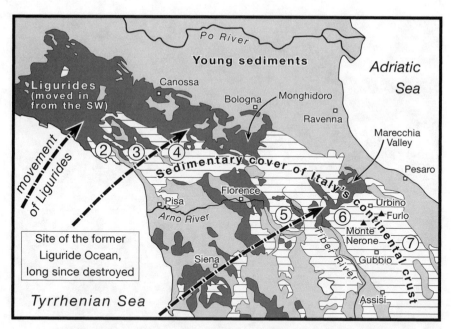

Rocks called Ligurides (dark gray) were driven out of the "Liguride Ocean" when that ocean closed up. Now they are found on top of the original sedimentary rocks deposited on Italy's continental crust (striped). Both are covered by still-young sedimentary rocks (light gray). This map is unavoidably messy because the Ligurides are messy. The circled numbers are Giovanni Merla's anticlines, described later in this chapter, bringing up deeper rocks and getting younger to the northeast.

Pennine Ocean are caught between the European continent below and the Italian continent above, but in the Apennines there is no sign of a continent driven over the Liguride oceanic rocks. Somehow the Ligurides have slid all the way across Italy, and we will explore their strange history later in this chapter.

In the process of traveling this distance of more than a hundred miles, the oceanic Liguride rocks have been badly damaged. Some parts have been so shattered and beaten up that they have turned into chaotic jumbles of rock fragments in a matrix of sheared, scrambled clay. The Italians were calling this debris the *argille scagliose*, the "scaly clays," Professor Maxwell told us.[4] The Ligurides are not entirely chaotic, however. In some places within the chaotic scaly clays, there are coherent blocks of rock of all sizes that have gone along for the ride.

Milly is standing by an outcrop of chaotic Liguride rocks, all full of torn-apart blocks of different sizes.

John Maxwell had studied one of the very large coherent blocks under the guidance of Professor Merla,[5] and he told us about it with the enthusiasm that geologists cannot suppress when they are talking about *really* interesting rocks. "It is at a place called Monghidoro, on the crest of the Apennines between Florence and Bologna," he said. "The Monghidoro block is ten kilometers wide and ten kilometers long, and about a kilometer thick. And you can tell from the graded bedding that the whole block is"—he paused for effect—"it's upside down!"

The Monghidoro slab is a huge block of rock, roughly the size and shape of the city of San Francisco and almost a mile thick.[6] How it got overturned is still not entirely clear,[7] but Professor Merla had come up with a remarkable explanation for how the Ligurides, including Monghidoro, got moved so far to the northeast over the top of the Italian continent. It was a theory a Hawaiian surfer would appreciate.

In addition to the great blocks of older rock caught up in the Liguride chaos, like Monghidoro, there are also younger sediments that were deposited on top of the moving Ligurides and have gone along for the ride. The Italian geologists call these hitchhiking rocks the Epiligurides, meaning "on top of the Ligurides," and they are scattered around he Apennines, mostly northwest of Florence.[8] Canossa, the castle of Countess Matilda of Tuscany, where the Emperor Henry IV stood in the snow, begging forgiveness of Pope Gregory VII, is built on a hill made of Epiliguride deposits.[9]

In his great synthesis of Apennine geology, Merla showed that there are seven or more anticlinal ridges, lying side by side in Tuscany, Umbria, and Marche.[10] Then he demonstrated that the ridges get younger and younger toward the northeast, using a variety of evidence—for example, the western ridges are more eroded, the eastern ones fresher, higher, and more continuous.[11] Merla's idea was that the Ligurides—the chaotic scaly clays with the coherent blocks—had *surfed* across the Italian Peninsula. As each underlying ridge rose, in sequential order, from southwest to northeast, the Ligurides would slide down the front side toward the northeast, and then the next ridge in turn would rise up, lifting the Ligurides again and allowing them to slide down ever farther east.

It was a wild, crazy story, and not all Italian geologists accepted it.[12] It was so outrageous that geologists from other countries just had to go and see the Apennines, and some stayed on to study and write about this remarkable place.[13] But notice how nicely it agreed with the preference of midcentury geologists for explanations that involved as little horizontal motion in the crust as possible. Through nearly vertical uplift of one anticline after another, in sequence, but no horizontal motion of the crust, gravity sliding allowed the Liguride sheets to move horizontally a hundred miles!

Although Giovanni Merla's ideas about the Apennines were attracting international attention in the 1950s and 1960s, it was a new generation of Italian geologists, growing up and living in these mountains, who really worked out what was going on.[14] The next great synthesis after Merla's 1951 study came in 1970, when the new generation of Florentine geologists devoted a whole issue of the journal *Sedimentary Geology* to the Northern Apennines.[15] In this set of major papers Professor Merla's former students laid the basis for our modern understanding of the Apennines. Most notably, in an astonishing piece of geological detective work, they pieced together the jigsaw puzzle of blocks floating in the Liguride scaly clays and showed that they were fragments of several quite different sequences of rocks ranging from Triassic to Eocene in age, and that each sequence had formed in a different part of the now-vanished Liguride Ocean.[16] Imagine working out the detailed geography of an ocean that was crushed shut and disappeared tens of millions of years ago!

The Florentine volume came out at a critical moment in geology. In 1970 the plate-tectonic concept was just emerging, and most geologists were not sure which way things would go. Would plate tectonics, with its heretical evocation of enormous horizontal motions, revolutionize geology, or was it a big mistake that would soon be corrected and go away?

In the third of a century since 1970, plate-tectonic theory has been abundantly confirmed, it has indeed revolutionized geology, and the old fixism has been swept away. Geologists no longer think that avoiding horizontal movements confers merit upon a theory of mountain building. Quite the contrary! Now, in a new synthesis of the Apennines, with studies by many authors,[17] the understanding of the Apennines is strongly informed by plate tectonics, there is lots of horizontal movement, but in a strange irony, the Apennines do not really fit into the usual understanding of plate tectonics. The difficulties will become apparent in chapters 12 and 14.

This part of Italy is reminiscent of the lower two-thirds of the tectonic sandwich of the Alps, where oceanic rocks are caught in a vertical pile between two blocks of continental crust, but there is no upper continental block over the Ligurides. Geologists today are trying to determine whether there used to be an upper continental block that has been completely eroded away, or whether there never was one.

I had the Ligurides in mind in the first couple of paragraphs of this book: "From high up on a peak called Monte Nerone, on clear, crisp autumn mornings, you can see far across the landscape of Italy. . . . And along the northern horizon stretches a wilderness of scrambled rock that has glided, mysteriously and imperceptibly, a hundred miles across Italy, hinting at the almost unthinkable wilderness of time that recedes back into our planet's past."

Thrusts and Folds, Sweeping Across Italy

And so we come now to the climactic event in the history of the Mountains of Saint Francis. As the front of the Earth storm passed through, the sedimentary rock layers that had lain so quietly for so long, accumulating on the sea floor, were squeezed into an array of great folds and emerged from the sea, bringing the Italian Peninsula into being. It was a complete transformation to a brand-new state, like the change of a caterpillar into a butterfly as it emerges from its cocoon.

In the years since Giovanni Merla made his great discovery, geologists have learned a lot about how folds form, in mountain belts all over the Earth. Some of our understanding has come from studying the rocks now exposed in deeply eroded mountain ranges. In addition we now have lots of information from deep wells and seismic exploration profiles in fold belts rich in oil, supplemented by the results of sophisticated computer models. Thus it is clear that folds form when beds of rock are compressed. Looking at outcrops in detail, we almost always find evidence that the beds have been shortened.[18]

If something, perhaps a colliding continent, is pushing some sedimentary rocks sideways, the moving rocks will slide on a weak layer. In our part of the Apennines there are three main weak layers—the Triassic Burano evaporites, the Cretaceous Fucoid marls, and the Miocene marls of the Schlier. The sliding takes place on a break called a thrust fault, and where the fault stays within a horizontal weak layer, we call it a "thrust flat."

Obviously the thrust cannot continue indefinitely as a flat surface down in the stack of sediments, because a place is needed for the displaced rocks to go. That space can only be at the Earth's surface, so the thrust has to cut upward toward the surface, usually at about thirty degrees. This upward-trending portion of the thrust is called a "ramp."

If you examine the next diagram carefully you can see how the wedge of moving rock above the ramp gets forced into the shape of an anticline. Geologists working in deeply eroded mountain belts or where there is a lot of seismic and well data have found that in regions of bedded sedimentary rocks, thrusts faults typically have this "ramp-flat geometry."[19] Apennine geologists are now pretty much in unanimous agreement that the folds in our mountains—Merla's ridges—are ramp anticlines.

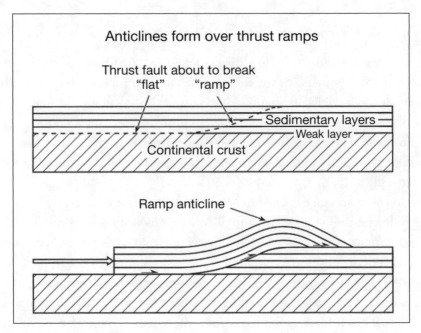

Anticlines form when the sedimentary cover is forced over a ramp, where a thrust fault cuts up from one flat level to another.

The diagram also shows that a ramp anticline has an asymmetrical internal structure—there is an inherent difference between its back and front sides. On the back side—the direction from which the thrust sheet is coming—the thrust lies far below the surface, but at the front of the fold, the thrust is much shallower, and may even reach the surface. This asymmetry is very clear where the Burano Valley cuts through the big anticline. It looks like the action in the anticline—the intense deformation—was all on the northeast flank, and the southwest flank has just gone along for the ride. Apennine anticlines tend to have gently folded sides on the southwest and strongly deformed, broken limbs on the northeast, where the ramp is exposed or only slightly below the surface.

A Whole Array of Folds, Forming the Apennines

The story of the folds is grander than a single anticline, however. The Mountains of Saint Francis are not limited to the one great fold where we find Monte Nerone, the Burano Gorge, Monte Acuto, and Monte

Catria. This part of the Apennines includes a whole range of large anticli-nal folds, from Monte Subasio in the southwest to Furlo in the northeast. Other, smaller anticlines continue on to the northeast, making up the lower hills near the coast. Still others, smaller yet, extend out under the Adriatic Sea, where the folds are still submerged beneath the water. We need to see how the concept of a single ramp anticline can be extended to explain an entire range of mountains. Again the understanding comes from other mountains that are more deeply eroded or have been more extensively drilled.[20]

In those well-understood fold mountains, geologists find a whole series of thrust ramps, each with its own anticline. These thrust arrays form pat-terns in which newer, younger folds form one after the other, in an orderly sequence, progressing in the same direction as the thrust mass is moving. It seems that after one ramp has been active for a while, building an anti-cline, it is hard for the displacement to continue on that ramp, where a huge mass of rock has to be pushed farther and farther. Instead the thrust slices forward along its original weak layer for some distance and breaks upward in a new ramp.

The new ramp builds a new anticline, and then the process repeats, with a jump to a still newer ramp, building another anticline, and so on. This produces a fault array below the surface, with a whole set of thrust flats and thrust ramps in a pattern that can be quite complex and chal-lenging to understand. At the surface itself, a set of parallel folds reflects the fault array, with each fold marking a ramp at depth. The folds get younger and younger in the direction the thrust sheets were moving.[21] Geologists use the term "propagating fold-thrust belts" for these features, which are now known all over the world.

This concept of a range of folded mountains forming, fold by fold, above an array of propagating thrust faults is just what we need to under-stand the Apennines. In some deeply eroded folds, like the big anticline in the Burano Valley, we can see from the amount of damage to the rocks that the ramp is closer to the surface on the northeast side and deeper on the southwest. So the fault array should be propagating from southwest to northeast and the anticlines should be progressively younger toward the northeast. That inferred decrease in ages toward the northeast beautifully explains the character of this part of the Apennines, where the anticlines get younger from Gubbio all the way out into the Adriatic Sea.[22] It is a fine explanation for Merla's migrating ridges, but of course this deep fault array is all an inference, because we can see only what is at the surface.

Listening to the Deep Structure of the Apennines

Geologists have long faced this frustration of not being able to look down into the Earth and see the deep structures that control what happens at the surface. What does the array of Apennine thrust faults look like at depth? If we could see those deep thrusts we would have a better idea of what is driving the great Apennine Earth storm.

We geologists can only envy astronomers who can actually *see* objects far across our galaxy and beyond. We cannot even see an inch below the Earth's surface, let alone kilometers down, where the critical action must be taking place. In this dilemma geologists have taken the advice of musicians—to really understand, close your eyes and listen.

Listening works for geologists because, although light cannot pass through solid rocks, *sound* can. Thus for mapping geologic features at depth geologists have come to rely on seismic exploration—a kind of musical technique.

In simplest outline the seismic method involves making a bang in the rock of the Earth's surface and listening for echoes. The bang may be the percussive drumbeat of a dynamite explosion, or we may use a sustained musical tone from a big, heavy vibrating pad. The sound passes down into the Earth, reflects off various surfaces—faults and contacts between formations—and comes back up to the surface, just as orchestral music reverberates off the walls of a concert hall. How long the sound takes to go down and come back up tells you how deep the reflector is.[23]

Oil company geologists routinely use the seismic method, which is expensive but has discovered the underground traps that produce much of our oil. Geologists from the Italian oil industry have explored the depths of the Apennines in this way, and have graciously shared many of their results.[24] It is harder for academic geologists to find the money for this tool, but in the 1990s a group led by my old friend Paolo Pialli was able to organize a seismic study, and they did find out what was under the Apennines.

Paolo was born slightly before I was; we are from the next generation after the great Italian geologists I have talked about so much. After
spending a day together in a car on a field trip in 1975, we became
lifelong friends. Paolo had been a student of Roberto Signorini at the
University of Rome. A new Institute of Geology was being founded at
the University of Perugia, and Roberto Colacicchi, a prominent expert
in limestone geology, moved there from Rome to start the geology program, bringing Paolo with him.[25] Together they did important studies of
the limestones that make the backbone of the Apennine ridges,[26] but
Paolo had another dream.

Italian geologists were among the best in the world at studies of sedimentary rocks, especially limestones and turbidite sandstones. But Italy
had few experts in structural geology, the study of deformed rocks. Paolo
was convinced that in order to understand the Apennines, Italian geologists had to master modern structural geology. I remember Paolo having
dinner by the fire at our house in Berkeley, on a visit just after he had
been promoted to full professor. He told Milly and me that he could now
work on whatever he found most interesting, and that he would be shift-

Paolo Pialli (seated) *with Bill Lowrie in the Apennines in 1976.*

ing to a focus on structural geology. Knowing that my doctoral work had been in that field, he said he would like to send the first member of his new group to work with me. And thus, a few months later, Giusy Lavecchia arrived at Berkeley as a postdoctoral fellow, and she was the start of Paolo's expanding structural geology group at Perugia.[27] Their work began a renaissance in structural geology studies of the rocks exposed in the Mountains of Saint Francis.

But Paolo knew that most of the answers lay hidden below the surface, and he worked tirelessly to start a program of seismic exploration of the Italian subsurface. Eventually the Crosta Profonda (CROP) program was launched, and Paolo and Tonino Decandia at Siena were the leaders for the CROP-03 profile that crossed the peninsula close to Siena and Perugia, passing through the Mountains of Saint Francis.[28]

Before the work of Paolo's group, many Apennine geologists thought that the folded sedimentary cover had just been pushed along, above the passive undeforming basement of the Italian continental crust.[29] Instead Paolo's seismic profile showed that five or ten kilometers down, thrust faults slice upward, out of the deep crust, and they have pushed parts of

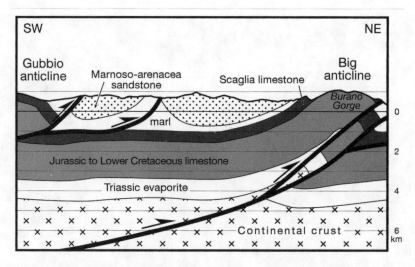

This interpretation of a deep seismic reflection profile by Paolo Pialli, Massimiliano Barchi, and the rest of Paolo's group at Perugia shows how thrust ramps have produced the anticlines and synclines in the figure on page 122. It also shows that the Apennine thrust faults (the heavy black lines) cut through the continental crust, so that in some places crust has been driven up over its sedimentary cover.

the crust over its own sedimentary cover. Still farther down it appears that really deep thrusts have pushed mantle rocks over the lower continental crust. To geologists this is a remarkable result. It shows that the Apennines are not just a superficial mountain belt. They involve the whole continental crust.

I remember at Princeton, long ago, one of the other grad students having a sudden insight. "The mantle is king!" he said. For us surface-oriented geologists just before the plate-tectonic revolution, unable to see deep into the Earth, it was a breakthrough in understanding. Today all geologists almost instinctively appreciate that we can only understand what happens at the Earth's surface by investigating what is going on at depth.

Paolo Pialli knew this, and the deep seismic studies he started have provided a major key to understanding the Apennines. Unfortunately Paolo is no longer with us to lead this effort, but his younger colleagues at Perugia, Massimiliano Barchi and Giorgio Minelli, and Giusy Lavecchia at Chieti, are carrying on his legacy.

And so the history of the Mountains of Saint Francis is falling into place, clarified by the concept of a propagating array of thrust faults and folds, formed in an environment where the rocks are being compressed all the way down through the continental crust.

Understanding the Apennines—Almost

The concept of a propagating thrust array gives us the tools to understand Giovanni Merla's great discovery. Merla showed that there are several great ridges, side by side, running along the Italian Peninsula between Florence and Rome, and he recognized that they get younger and younger toward the northeast.

Merla could not have known why the ridges formed one after the other, because the concepts of ramp anticlines and propagating fold-thrust belts had not yet been discovered. But those concepts now let us understand that the Apennines are the solidified waves of a storm of geological deformation that swept eastward, and may *still* be sweeping eastward, converting quiet sea floor into mountain ridges that last a few tens of millions of years before erosion wears them away.

These concepts also let us understand why each turbidite unit was driven eastward over the next younger set of turbidites—the intriguing pattern that emerged in the previous chapter. We now realize that the weight of a new anticline pushes down the crust in front of it, shaping a depression on the sea floor. The depression channels and traps the dense turbidity flows. As the thrust array and its folds have advanced toward the northeast, a whole succession of turbidite sandstone formations has been deposited, each marking the position of the thrust front at one stage in its history. As each thrust in turn breaks its new ramp up to the surface, it encounters the newly deposited turbidites, and they get overridden by the ramp anticline, which includes the older turbidites. In this way the *Cervarola* turbidites were driven over the slightly younger *Marnoso-arenacea* turbidites. It was this thrusting that tricked Guido Bonarelli into getting the wrong age for the *Marnoso-arenacea* turbidites, but now we know the trick and it provides one of our keys for understanding the ridges that form the backbone of the Mountains of Saint Francis.

The Italian Peninsula exists because of Giovanni Merla's ridges. If they had never formed, all this part of Italy would still lie submerged below sea level. Human history would have unfolded in an entirely different way—with no Rome, no Roman Empire, no Gothic War, no Roman Catholic Church, no Saint Francis, no Florentine Renaissance, no Tuscany for Nicolaus Steno to invent geology in, and a greatly impoverished human treasure of art and literature, music and science. Our civilizations are built on the landscapes that geological history has constructed.

It all seems to make sense, but there is one major complication left to explore. We now know that parts of the compressional Apennines are no longer in compression, but just the opposite—they are being extended and pulled apart. This is not at all what most geologists would have expected, and it is the final part of the mystery of the Mountains of Saint Francis. What could be tearing the Apennines apart?

12

TEARING THE APENNINES APART

ASSISI SLEPT PEACEFULLY in the small hours of the night of September 26, 1997. Suddenly, with no warning, the medieval buildings were moving—back and forth, up and down. Centered beneath nearby Colfiorito, a magnitude 5.6 earthquake was violently shaking this part of Italy. Roofs and walls, centuries old, were severely damaged, and people ran in panic into the streets. Then stillness returned to the stricken city.

Ten hours later, as officials and a television crew were inspecting the Basilica of San Francesco for damage, a second, stronger earthquake struck, with a magnitude of 6.0. Frescoes came loose from the ceiling as part of the roof of the basilica collapsed and fell in a cloud of dust. Two engineers and two Franciscan brothers died in the debris. People were left homeless all over this part of Umbria.

Why did this earthquake strike Assisi?

The answer is not at all obvious from what we know about plate tectonics. Some big earthquakes happen along subduction zones, where an oceanic plate is slipping, not so gently, beneath an overriding plate. This is the case west of Sumatra, where a subduction zone marked by earthquakes on a plane dipping deep into the mantle was responsible for the terrible earthquake and tsunami of December 2004, but there is no belt of earthquakes marking a subduction zone near Assisi. Other big earthquakes occur along transform faults, where one plate is sliding horizontally past another. This was the explanation for the great San Francisco earthquake of 1906, but there is no transform fault in this part of Italy.

The Assisi earthquake came from something different, and it offers us a clue to the origin of this unusual mountain range. Let us approach the question of the Assisi earthquake in a roundabout way, by looking closely at the landscape.

Observing the Landscape

Back at the very beginning of this book, in the account of our Christmas visit to Assisi in 1970, I described how puzzled I was by the broad, flat valley of the Tiber River we followed on our way to Assisi, far from the sea, in a place where I would have expected deep, rugged canyons. Nothing we have seen so far can account for the mystery of the flat valleys, but we are getting close to an explanation.

The answer to the flat valleys and to the Assisi earthquake will begin to emerge if we examine the landscape of this part of Italy with more care than we have done so far. Geologists today have new tools for studying landscapes, and one of them is the digital elevation model. To make a digital elevation model, an area of interest is divided up into little squares, and the average elevation for each square is either read off a topographic map or measured directly by radar from space.[1] The smaller the squares, the higher the resolution of the digital elevation model, and the more detail you can see.

A digital elevation model can be displayed as a map in various ways. One way is to show higher elevations in lighter tints, as in the shaded relief map of the Italian Peninsula from the Po Plain to Rome on page 202.

On the image of the topography we can see the irregular west coast of the peninsula, facing the Tyrrhenian Sea, where the large island of Elba—the temporary exile of Napoleon—looks like a fish swimming westward, accompanied by a school of smaller islands. The eastern coast, facing the Adriatic Sea, is much smoother, interrupted only by the headland of Monte Conero, site of the city of Ancona—a name appropriately derived from the Greek word for "elbow."

The relief map shows that the highest part of the peninsula lies closer to the Adriatic side. In the southeast corner are the very high Apennines of the Region of Abruzzi, east of Rome. From there, the high country, marking the border between Umbria and Marche, trends northwest to the north center of the image, east of Florence. West of Ancona lies the portion of Apennines I like to call the Mountains of Saint Francis, formed by a set of parallel ridges trending north and northwest. These are the anticlinal folds, formed above thrust ramps that were the focus of the previous chapter.

The western half of the peninsula lies at lower elevations and has a considerably more irregular geography. In this area, comprising Tuscany, the western part of Umbria, and Latium, a number of valleys seem to wander around in a rather disorganized pattern. The major rivers are influenced by these valleys but do not follow them slavishly. In the northwest corner, the Arno reaches the Tyrrhenian Sea near Pisa, and upstream it cuts through various highs and lows, including the oblong valley where Florence straddles the river. East of Florence the Arno loops around in an unusual way.

Siena's river, the Ombrone, reaches the Tyrrhenian coast east of Elba. Finally the Tiber stretches upstream from Rome, passing east of the three big craters that mark the Roman Volcanic Province, and its headwaters drain the Umbria-Marche Apennines. The small, high crest of Monte Subasio marks the site of Assisi, looking down on the broad valley that surprised me in 1970.

Compared with the long, irregular routes of the three large Tyrrhenian rivers, there are many smaller rivers flowing into the Adriatic, but they are short, fairly straight, and have few large tributaries, so most of the drainage of this part of Italy flows west into the Tyrrhenian Sea. This river pattern has been familiar to people living on the Italian Peninsula as far back as Roman or Etruscan time, but only recently has there been a geological explanation, and it is a remarkable one.

Roberto Signorini and the Tearing Apart of the Apennines

In my mind the explanation of the Apennine valleys is closely associated with two more of the great names in Italian geology—Roberto Signorini and Livio Trevisan—contemporaries of Carlo Migliorini and

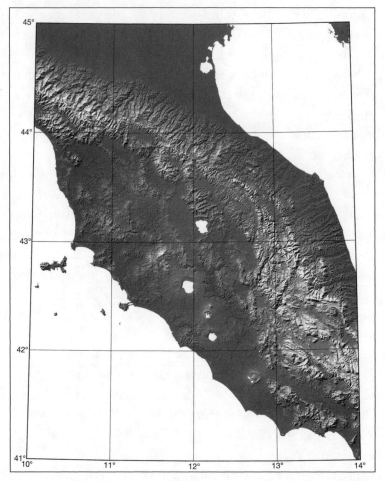

Relief map of the northern Apennines, made with the new elevation data from the Shuttle Radar Topography Mission. The higher Apennines cross from the west coast to the east around latitude 44° N, then curve south. Tuscany is the area of valleys and lower hills within the arc of the mountains.

Giovanni Merla in the generation that was active just before and after World War II.

Roberto Signorini, descended from a family of Florentine artists, spent the first twenty years of his career with the Italian petroleum company, AGIP. Signorini was thus familiar with all the seismic data and oil wells that showed what was happening deep beneath the surface.

In addition he spent his free time in the mountains doing geology for the pure love of it, and made many fine discoveries. It was the heroic age

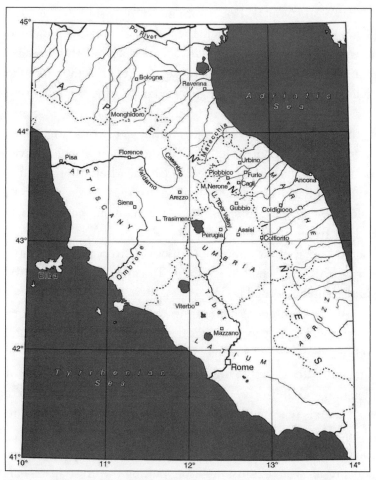

Geographic map of the same area as the relief map of the preceding figure. Note especially the difference between the rivers on the Tyrrhenian side—few, large, and winding—and those on the Adriatic side—many, small, and straight. This is an important clue.

of Italian geology, when geologists as yet knew little about the rocks of the Apennines. Signorini's short notes on his observations gave the first reports of many discoveries that today provide the focus for whole groups of researchers. He was the first to describe two of the key Apennine features that have been important in this book already—the great upside-down slab of turbidites at Monghidoro that John Maxwell told us about in class at Princeton, and the progressive displacement of the turbidite basins toward the northwest that we now know marks the migration of the compressional front.[2]

During World War II, Roberto Signorini was stationed at the AGIP headquarters in Rome, and he held the Italian Geological Society together in the terrible days of the German occupation and the passage of the Allied armies northward up the peninsula. To me it is moving to read the proceedings of the Società Geologica Italiana during the war years and to see how geologists like Signorini did their best to keep their science alive during the social and personal tragedies that were unfolding.

After the war Signorini left AGIP and taught at the University of Rome until 1968, when he moved to Siena, where a university dating back to the early thirteenth century had just recently created an Institute of Geology. Signorini thus spent his last years doing geology back in his native Tuscany.

There is yet another remarkable discovery in one of Signorini's short notes from the war years.[3] In southern Tuscany he found that the rocks were broken up by normal faults.

At first this sentence, containing the word "normal," seems quite unexceptional. But not to a geologist, who would read the end of the last sentence as, "broken up by *normal* faults!"

Remember from the previous chapter that "thrust faults" result from compression, and they drive older rocks up over younger ones, so that the rock sequence is repeated below and above the fault, and the whole pile of rocks gets thicker.

"Normal fault," as a technical term, means just the opposite. Normal faults result from extension—from pulling the crust apart—and they drop younger rocks down over older ones. Instead of thickening the rock sequence, normal faults make the rock sequence thinner.

And so for Signorini it was a great surprise to find, in the compressional belt of the Apennine folds, the extension represented by normal faults. He was the first geologist to glimpse the unexpected combination of compression and extension that is fundamental to the mystery of the Apennines.

I really wish I had been able to know

Roberto Signorini (1901–80).

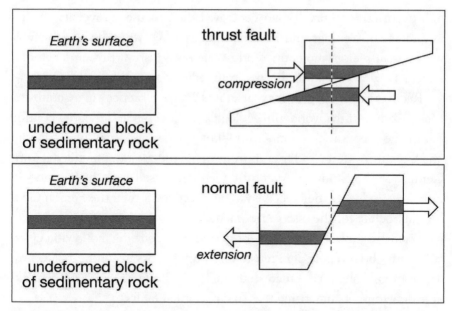

Thrust faults form where the Earth's crust is compressed. Note how the block gets thickened and the gray layer is repeated along the dashed line. Normal faults form where the crust is extended. Notice how the block gets thinner and the gray layer is missing along the dashed line.

Roberto Signorini. Everything I have heard from friends who did indicates that he was not only a great geologist but a true gentleman as well. "He taught us how it is possible to combine the duties and culture of a great scientist with the kindness and modesty of a simple and decent person."[4]

Livio Trevisan, Earth-Storm Waves, and the Apennine Valleys

I did have a chance to meet Livio Trevisan, completely by accident, while Bill Lowrie and I were having dinner one night in 1976 in Gubbio, after a day of fieldwork. Geologists do not have much trouble recognizing one another, and we ended up spending the evening with Professor Trevisan and his younger colleague Renzo Mazzanti. They were also in Gubbio to do field research, and the paper that resulted from their work has had a big influence on my thinking about the Apennines.

We have already met Livio Trevisan as the author of the little cartoon about gravity sliding. Professor Trevisan was one of the giants of mid-twentieth-century Italian geology, called to the University of Pisa as director of Geology and Paleontology in 1939, when he was thirty. It was an inauspicious moment; World War II was just starting, and before it ended, Trevisan would see his institute reduced to rubble.

Recovering in the postwar years, Pisa became Florence's main competitor as a source of the wonderful geological discoveries and theories coming out of the Apennines. Competitor—but partner also, for Livio Trevisan and Giovanni Merla, old friends from their student days, founded a joint Center for the Study of Apennine Geology, based both in Pisa and in Florence, and later they received a joint recognition by the French Geological Society for the discoveries made by the geologists of the center.[5]

Livio Trevisan seemed to be interested in everything. Following up on Signorini's brief report of extensional normal faults in southern Tuscany, Trevisan published a detailed study in 1951, illustrated with his remarkable drawings, documenting the Tuscan normal faults.[6]

And then, in 1975, Trevisan and his friends produced a new bombshell paper that has strongly shaped geological thinking about the Apennines ever since and has major implications for our global understanding of how the Earth works.[7] In this remarkable study the Pisan geologists showed that the extension in Tuscany has *migrated* northeastward across Italy. Thus, in addition to the compressional front moving gradually northeastward, which geologists had known about since the work of Signorini and Merla, there is an extensional front following along behind it, about a hundred kilometers in the rear. Geologists don't usually think of it this way, but

these migrating fronts of extension and compression are like the waves of the huge Earth storm that is sweeping, ever so slowly, across Italy. This kind of behavior had never been recognized anywhere else in the world—just the sort of surprise we have come to expect from the Apennines!

These coupled fronts, compressional and extensional, migrating in lockstep across Italy, imply that the Apennines are forming in a

Livio Trevisan (1909–95).

Professor Trevisan took this photograph of the Geology and Mineralogy Building at the University of Pisa, destroyed by a wartime bombing raid.

very strange way. At the compressional front, sedimentary layers that have been accumulating quietly for more than 150 million years suddenly receive a flood of turbidites and then get thrust-faulted into an anticlinal fold. The anticline spends several million years as part of a fold-thrust belt while the compressional front moves farther on and the extensional front approaches. When the extensional front arrives, the fold is torn apart by normal faulting, and subsides in elevation as the crust beneath is stretched and thinned. Geologists are still working out the implications of this strange kind of behavior.

In an extensional regime, normal faults break the crust up into blocks. Uplifted blocks are attacked by erosion, and the erosional debris accumulates in the low blocks. This is the nature of the gentle and picturesque Tuscan landscape—it is a patchwork of hills made by uplifted and partly eroded blocks, separated by valleys that represent the partly filled low blocks in between.

Toward the northeast, in Umbria, where the normal faulting was more recent, the uplifted blocks are higher, less eroded, and more continuous. The low blocks also are better defined, and they make long, fairly straight valleys, clearly visible on the relief map. The most prominent of these long valleys are, first, the upper Arno Valley—the Valdarno—with continu-

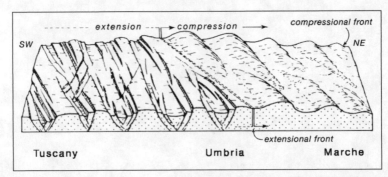

Livio Trevisan drew this sketch to show the extensional front following behind the compressional front as they both move toward the northeast, generating the Apennines.

ations to the northwest beyond Florence and southeast toward Lake Trasimeno, and second, the upper Tiber Valley that runs northwest from Perugia, with a branch that runs southeast, past Assisi.

Past Assisi! This is the wide, flat valley that you look down on from the city of Saint Francis—the valley that I found so odd as we first approached Assisi in the winter of 1970, and the valley where the rainbow touched down on the dome of the Basilica of Santa Maria degli Angeli. At last we have our answer! This valley is a fault block that dropped down when the migrating extensional front passed through. It was *indeed* odd to find a broad flat valley in the midst of a young mountain range. But the Apennines are an odd mountain range, where the unusual interplay of compression and extension has created an unusual landscape.

An Evolving Landscape

The 1975 paper of the Pisan geologists had barely been published when Livio Trevisan was out in the mountains with Renzo Mazzanti, trying to see what the coupled fronts implied for the origin of the Italian landscape. That was the main thing we talked about during our chance dinner at Gubbio. A couple of years later he sent me the remarkable paper in which they presented their results.[8]

The paper by Mazzanti and Trevisan was an attempt to explain the mystery of the Apennine rivers that you can see in the relief map at the beginning of this chapter.[9] Why are the Adriatic rivers short, straight, and numerous, while the Tyrrhenian rivers are long and winding, with only three major ones between Florence and Rome?

The Tyrrhenian rivers quickly made sense in light of the extensional front. Whatever arrangement of rivers there used to be, the passage of the extensional front would break the landscape up into a pattern of elongated fault blocks. This pattern would make easy routes for the rivers, which would follow a low block as far as possible and then cut over into the next one. This explains the zigzag pattern of the Arno, the Ombrone, and the Tiber. The Tiber, for example, has two long tracts where it flows from northwest to southeast along down-dropped blocks, and in between it cuts across the grain toward the southwest.

The Arno follows an even stranger path through Tuscany. It flows southeast out of the valley called the Casentino, almost reaches Arezzo, but makes a complete U-turn and heads northwest to Florence and then down to the Tyrrhenian Sea. Dante Alighieri, around 1300, was aware of this strange route, and used it to insult Tuscany in the *Divine Comedy*. He has Guido del Duca, a man with a pathological dislike of Tuscans, describe the Arno as an "accursed, wretched ditch." But even such a disreputable river, says Guido, when it approaches the Tuscan town of Arezzo, is so disgusted that it "averts its snout" and finds another path to the sea![10]

The Adriatic rivers are much harder to explain.[11] Why do they flow straight down to the Adriatic, chopping right through big anticlinal mountains that ought to have been barriers to them? The gorge at Furlo, which we saw from the top of Monte Nerone at the beginning of the book, is just one spectacular example of how the Adriatic rivers pay no attention to the mountains, and simply cut right through them.[12]

Furlo and the other gorges of the Adriatic rivers presented a great mystery, but Mazzanti and Trevisan proposed a solution that I find extremely elegant. In the skillful drawings for which Professor Trevisan was known, you can follow the sequence of events that produces a gorge like Furlo. The drawings do not show the internal thrust structure, for that was being worked out at the time by other geologists, but they correctly show the anticlinal ridges growing one after the other as the compressional front migrates toward the Adriatic Sea on the right.

As each ridge begins to rise from the sea floor, it forms a sediment trap,

The Furlo Gorge, where a river slices through one of the limestone ridges of the Mountains of Saint Francis. The Roman road called the Via Flaminia passes through the Furlo Gorge.

behind which the sand and silt from a river mouth gets deposited. That sediment builds up to sea level and makes a new coastal plain, allowing the river to lengthen in the downstream direction and thus to be in place just as the next ridge rises above sea level. This allows the river to begin cutting its canyon, and to deepen it as the ridge continues to rise.

I always loved this beautiful explanation of the deep gorges, but it was a simplified diagram in a brief paper in an obscure journal, and it didn't seem to me that their work had received the recognition it deserved. Many years later, trying to get their ideas more widely known, I wrote a longer paper that tested the Mazzanti-Trevisan hypothesis carefully, found it to be in good agreement with the actual geology, and expanded the ideas that they had put forth.[13]

Deep Discovery

From our study of the Italian landscape we have seen that the flat valley floor at Santa Maria degli Angeli, below Assisi, is a down-dropped fault block. It sank down because of the extensional tearing apart of the slightly older compressional folds of the Apennines. While sinking, it has been partly filled by sediments eroded from the surrounding hills and mountains.

Does this sound familiar? This is just what we talked about in chapter 5 (see especially the figure on page 79) as a modern version of the giant, collapsed caverns that Nicolaus Steno invoked to explain the flat-bottomed valleys of Tuscany. Steno's caverns, which would have

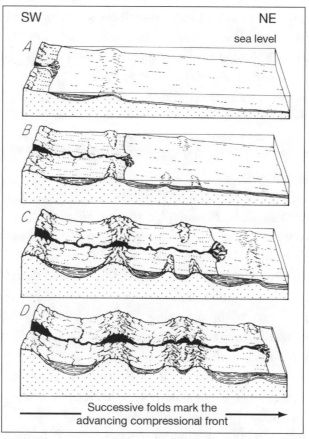

This drawing, modified from Mazzanti and Trevisan's 1978 paper, shows how the Adriatic rivers of Italy may have been able to cut through mountains and excavate deep gorges like Furlo. The landscape is gradually evolving, from sketch A to sketch D.

opened up during extensional normal faulting if rocks had been strong enough, never did open, because they collapsed even as the extension was occurring. Nevertheless Steno's deep insight into the history of the Earth really amazes me, and it is even more astonishing because no one had ever thought this way before—Steno was inventing geology as he went along.

Geologists now know that sediment-filled valleys in extended regions form above curving listric normal faults that flatten out at deeper levels. So the obvious question about the valley below Assisi is: Which way does the listric fault dip? Does it dip and flatten toward the southwest, or toward the northeast?

Geologic mapping has shown that there is indeed a fault between Assisi and Santa Maria degli Angeli, with the valley side dropped down, and the fault does dip southwest.[14] That fault could either be the listric normal fault itself, if the listric fault dips southwest. Or it could be one of the little adjustment faults, if the listric fault dips northeast. I think most geologists would have expected the fault to dip southwest, and I certainly did. Looking at the situation in a broad context, it would make sense for the fault to flatten out toward the southwest, back toward Tuscany, which had previously been extended. However, to our surprise, the fault actually dips in the opposite direction.

The correct picture came from the seismic reflection profile that Paolo Pialli and his Perugia team made. On that profile of the deep crust, they could see very clear seismic reflections coming back from a gently dipping normal fault responsible for the valley of the Upper Tiber.

I believe it was Massimiliano Barchi and Arnoud de Feyter, studying the brand-new seismic profile in 1995, who first noticed that the master fault seemed to dip the wrong way. The fault does *not* dip southwest toward the Tyrrhenian Sea. It actually dips in the other direction, northeastward toward the Adriatic. This master fault of Apennine extension came to be known as the Altotiberina Fault, after the valley of the upper Tiber River. The normal fault at Assisi, it turned out, is not the main fault, but just a minor adjustment fault dipping in the opposite direction. It was a major discovery, and the rest of Paolo's group joined in to test the interpretation and figure out what it meant.

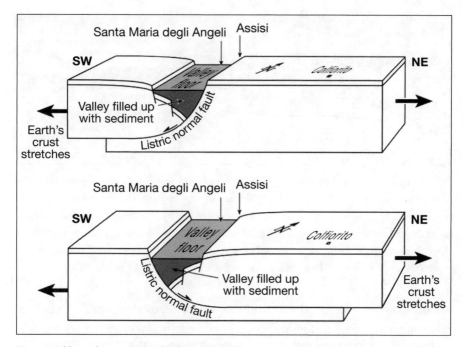

Two possible explanations for the flat valley of Santa Maria degli Angeli below Assisi. In the upper sketch the valley is due to a spoon-shaped listric normal fault dipping southwest. In the lower sketch the listric normal fault dips the other way. Most geologists would have expected the upper sketch to be correct, but this turned out not to be the case.

They published a series of papers[15] culminating in a full-scale synthesis establishing the reality and fundamental importance of this strange and unexpected fault.[16]

The timing of those studies was remarkable. Just as the geologists at Perugia were studying the completely unexpected Altotiberina Fault, the two 1997 Colfiorito earthquakes devastated nearby Assisi. A few years earlier the Colfiorito earthquakes would have been an unexplainable geological puzzle, just as historical earthquakes in this area had always been.

But the Perugia group now knew exactly what was going on. With seismological techniques, geologists could tell that this was an extensional earthquake, not a compressional one. The Colfiorito earthquake was due to extensional slip on the newly discovered Altotiberina Fault, slanting

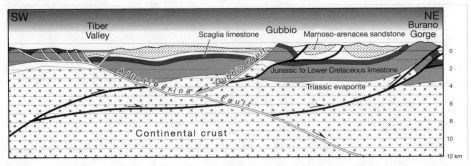

This cross section, based on seismic studies, expands and deepens of the earlier profile of the figures on pages 122 and 196. Here we see that the thrust faults (thick black lines), which compressed and shortened the continental crust, are cut by younger normal faults (heavy lines with white centers) that have extended and pulled apart the crust. These normal faults have dropped down the areas, now filled with sediments, that make up the Tiber Valley, the valley west of Gubbio, and the valley (not on the line of this profile) below Assisi.

gently down to the northeast and passing beneath Assisi and Colfiorito. During the intense follow-up study of the earthquake, Paolo's group was able to understand it in this new geological context.[17]

So now we realize that the Mountains of Saint Francis are the result of a complicated history. First a wave of compression has passed through, accompanied by a flood of turbidite sands and generating thrust faults that have driven up ramp-anticline folds. A wave of extension is following along behind, tearing apart the folded Apennines, dropping down fault-block valleys that fill with sediment, and completely rearranging the pattern of rivers.

In this view of the Apennines we can see a close geologic similarity between Assisi and Gubbio. Each town sits on the western edge of a mountain—Monte Subasio at Assisi and Monte Ingino at Gubbio. Both of those mountains are anticlinal folds that formed above thrust ramps as the compressional front passed through, about 10 to 15 million years ago. And both of those folds were cut in half by normal faults as the extensional front passed through in the last few million years. So Monte Subasio and Monte Ingino are both "half-anticlines," with the eastern half still standing high and the western half dropped down to form the valleys that lie just west of each town. Gubbio is the site of the easternmost normal fault, and thus it marks the present position of the migrating extensional front.

We now have a clear idea what has happened to produce the Apennines. They have resulted from a surprising combination of compression and extension. All that remains is to try to understand what motions, down in the mantle, could be producing this odd behavior. We need just one more key to unlock the mystery, and so we must now pause and look at a truly astonishing event in Earth history that has left an unmistakable mark in Italy.

13

SALT CRISIS

SOMETIMES THE MOST amazing events in the history of the Earth are first glimpsed through subtle, almost imperceptible hints. Like this one . . . In the Mountains of Saint Francis, close to the route of our winter trip of 1970, the usual marls and turbidite sandstones of the Miocene and Pliocene are interrupted by a Messinian interval in which there is a lot of gypsum.[1] There is Messinian gypsum also in the extensional grabens of Tuscany.[2]

The Messinian is the last stage in the Miocene. It is named after the Strait of Messina, which separates Italy from Sicily, where the Late Miocene also contains halite, or rock salt. The Messinian represents the time from roughly seven to five million years ago.

The Messinian gypsum and halite seemed a small thing, and the early geologists did not find them particularly exotic. But they provided the first hint of an astonishing story and a prelude to one of the great geological discoveries of all time—the Messinian salinity crisis of the Mediterranean.

Gypsum in itself is not strange. It is familiar in everyday life as the filling of wallboard used in building construction, and as the solid white material that forms when plaster of Paris is wetted and allowed to dry. In nature gypsum occurs as beautiful, shiny, clear crystals so soft they can be scratched with a fingernail. Gypsum and halite crystals grow when seawater evaporates, becoming so concentrated and saline that the dissolved salts begin to crystallize. For this reason halite and gypsum are called "evaporite" minerals.[3]

Seawater can evaporate in shallow lagoons along hot arid coasts, and gypsum is quite a common sedimentary rock. So at first the Messinian gypsum in the Apennines did not attract a great deal of attention. . . .

The Secret Canyon of the Nile

On the other hand, here is something *really* strange, although at first it did not attract much attention either, because almost no one knew about it. In Egypt, far to the east of Italy, out of sight beneath the Nile River, a monster canyon lies hidden, completely filled in with sediment. At first there was no reason to think it has anything to do with the Apennine gypsum, but now we know that they are intimately related.

The Nile has always been a river of mystery. Here comes this great river, flowing northward across a thousand miles of hyperarid Sahara Desert to the Mediterranean Sea, from an unknown source somewhere far to the south. And each year until the Aswan Dam was built, the river would flood Lower Egypt during the pitiless heat of September, carrying down a huge load of new silt that would yield rich crops to the Egyptian farmers. Where does the Nile come from?

"For two thousand years at least the problem was debated and remained unsolved; every expedition that was sent up the river from Egypt returned defeated."[4] Explorers who tried to follow the river upstream to find its source, driven by fanatical curiosity and courage, endured terrible hardships, especially where the Nile lost itself in the maze of impenetrable swamps in central Sudan called the Sudd.

Geologically also the Nile is a tangle of enigmas.[5] Even after the source of the river was finally recognized in the great lakes of high, rain-drenched East Africa, the Nile continued to guard its further mysteries.

In 1960 Egypt began the construction of the Aswan High Dam, with help from the USSR. At the Aswan site, five hundred miles before it reaches the Mediterranean, the Nile flows through a valley cut into ancient granite. There is river silt on the floor of the valley, and I imagine the engineers expected the weak silt layer to be thin, so the foundations of

The Aswan Dam under construction in 1966, when only Russian and Egyptian geologists knew about the deep, buried mystery canyon beneath it.

the dam could easily be driven into the strong granite beneath. But when the engineering geologists drilled down into the silt to find out how thick it was, they received an unwelcome surprise.

You would certainly not expect the fill of young sediments to go deeper than sea level, a hundred meters below the Nile at Aswan, because a river stops flowing when it reaches the sea and cannot cut deeper than that level. But the drill bits kept on going down, below sea level, still in young sediments deposited by the Nile. And then, about forty meters below sea level, the nature of the sediment fill changed. No longer were the drillers cutting through river sediments, but now in the deeper cores, they were finding sediments containing *marine* fossils!

What were marine sediments doing here, buried in an ancient Nile gorge five hundred miles from the Mediterranean Sea? How could the sea have ever been here, in the middle of the Sahara Desert, and in a geologically very recent time? You can almost picture the Russian and Egyptian geologists, scratching their heads in amazement.

At Aswan, the deepest part of the mystery canyon of the Nile extends down about three hundred meters—almost one thousand feet—below the present Nile River, and almost two hundred meters below modern sea level. For the engineers it raised the problem of how to build a safe dam when water can seep through soft sediments underneath it. For the geologists it presented a frustrating mystery. A river can only erode a canyon if there is some deeper place for the water to run down to. Aswan is only a hundred meters above sea level, and the sea is five hundred miles away to the north. The gradient of the Nile is thus very gentle, and there is no lower place for the river to go to. How could the river possibly have eroded a canyon in this part of Egypt?[6]

Yet the story of the Nile is even stranger, as other geologists discovered. Far to the north of Aswan, the Nile spreads out to form its great, fertile delta between Cairo and the Mediterranean Sea. Beneath the Nile Delta, just as at Aswan, there is a buried canyon, but this one dwarfs the filled-in channel beneath the dam! It seems clear that a huge canyon underlies the entire Nile, at least as far upstream as Aswan, and the canyon gets deeper and deeper downstream to the north. This canyon, too, was unknown, until it was discovered by petroleum geologists assessing the oil potential of the sediments beneath the delta.[7] Oil exploration wells brought up samples, so the geologists knew what kind of rocks, of what ages, lie far beneath the surface. Using seismic techniques, by bouncing sound waves off the rock layers at depth, they constructed a map of the subsurface.

The seismic data and the well samples showed that there had already been a Nile Delta back in the middle Miocene. To a geologist a

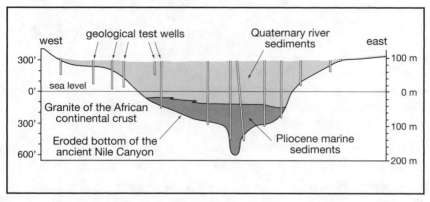

The sediment-filled canyon cut into the granite of the African continental crust at the site of the Aswan Dam, on the Nile. Note how much this diagram resembles the profile in Rome that Albert Ammerman and I made of the ancient Tiber Valley inside the Capitoline Hill, also using information from wells (see page 53).

A satellite image of Egypt. The Nile River carries water from the highlands of East Africa northward across the Sahara Desert to the Mediterranean Sea. The Nile Valley from Aswan to the delta is the site of an enormous ancient canyon, now filled with river sediment.

delta is not just the flat land above sea level; it is a huge body of sediment carried by a river and built out over the ocean floor, with only its very top above water. At some point the Nile River suddenly cut down deep into that mass of delta sediments, carving a great canyon reaching down to 4,500 meters—2.5 miles—below present sea level. Today that would be impossible, because rivers can cut down no farther than sea level. Something must have been different when the canyon was eroded.

It began to look, to those few geologists who knew about this information, as if sea level in the Mediterranean had gone way down for some
interval of time. That would explain the erosion of the great canyon
beneath the Nile Delta. If sea level were to fall, the water in the river
would have a lower place to go, and the river would be able to cut a deep
canyon. Not only would that explain the great canyon beneath the delta,
but it would also account for the canyon at Aswan, because when rivers
cut down, they also erode back upstream. But why would sea level drop so
far and then come back up? No geologist could imagine such a thing. It
was completely mystifying.

When had the Nile canyon been cut? In the oil wells of the delta and
the engineering wells at Aswan, the answer was the same. In both cases
the marine sediment immediately overlying the erosion surface dated from
the very beginning of the Pliocene, 5.3 million years ago,[8] so the erosion
must have occurred in the Late Miocene—in the Messinian. If indeed the

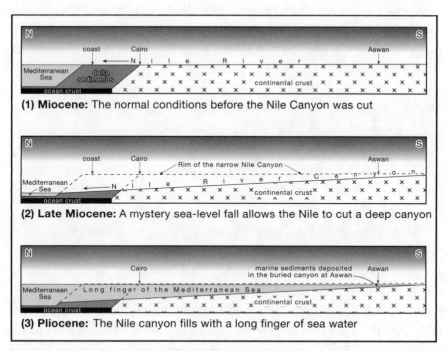

(1) **Miocene:** The normal conditions before the Nile Canyon was cut

(2) **Late Miocene:** A mystery sea-level fall allows the Nile to cut a deep canyon

(3) **Pliocene:** The Nile canyon fills with a long finger of sea water

*Sea level in the Mediterranean would have had to drop briefly, by a couple of miles, to allow
the Nile to erode the deep buried canyon at Aswan and beneath the Nile Delta. Here are the
steps that seemed to be required. No one had ever heard anything like this.*

sea level of the Mediterranean had been low when the canyon was cut, it must have risen very suddenly, because the Russian wells found marine fossils in the base of the Pliocene sediments that fill the canyon at Aswan.[9]

Imagine the great Nile canyon, suddenly flooded by the sea, so that a long, thin arm of the Mediterranean extended hundreds of miles upstream from the coast! Had there been oceangoing ships in those prehuman times, they could have steamed all the way to Aswan. There is nothing remotely like that finger of ocean anywhere in the geography we know today. It must have seemed—to the few geologists who knew about it—that there was a great story here, waiting to be understood.

Unlucky Leg 13

The geologists who finally understood the story did not even know about the Nile mystery. And they were not looking at rocks on land, as geologists had always done. They were using a brand-new geological tool—the deep-sea drilling ship, *Glomar Challenger*. In the 1960s, geologists had begun to realize that there must be a treasure trove of information about Earth history in the sediments on the ocean floor, where there are no waves and almost no currents to disturb the record. But the only way to get at that information was to take drill cores of the sediments. The Deep-Sea Drilling Project was developed to do this coring, and many difficult technical problems had to be solved before the *Challenger* could begin to sample the deep-sea historical record. Drilling began in 1968, and for more than thirty-five years the program has been a phenomenal success, certainly one of the most productive big-science efforts ever undertaken.[10]

In August 1970 the *Challenger* sailed out from Lisbon on the thirteenth leg of its voyage of exploration. Leading the expedition were two young scientists—Bill Ryan of Columbia University's Lamont-Doherty Geological Observatory, where we would become good friends when I arrived the following year, and Ken Hsü, a Chinese-American geologist who was a professor at ETH, the prestigious technical university at Zürich, in Switzerland. On board was an international team of geologists capable of determining the age of sediments based on the microfossils they contain, and of recognizing their environment of deposition.[11]

Glomar Challenger, *the ship of the Deep-Sea Drilling Project that began the effort to read Earth history in the sediments on the ocean floor.*

Bill Ryan had already been on four oceanographic cruises in the Mediterranean, collecting seismic reflection profiles to study the layering in the sediments below the sea floor. Bill was intrigued by a prominent level in the sediments that reflected sound waves back very strongly. He called it the "M-reflector." He had traced it all over the Mediterranean, but oceanographers found it nowhere else, and no one knew what kind of rock it represented or what its significance was. The M-reflector was a major target of Leg 13. Bill Ryan and Ken Hsü were determined to find out what it was.

It was not easy to find out. Ken Hsü later wrote a short popular book about this first drilling in the Mediterranean,[12] and it tells a harrowing story of the obstacles and frustrations they encountered. He and Bill had to face problems with navigation, with stuck drill bits, and with cores that came up empty. They had a continuing saga trying to connect with one of the scientists, who chased all over southern Europe before he could finally get ferried out to the ship. Scientists are not normally troubled by superstition, but on the *Challenger* they finally began to wonder whether it was simply unlucky to be on Leg 13!

Through all their difficulties Bill and Ken had to make agonizing decisions—whether to keep drilling a problem-infested hole or move on to

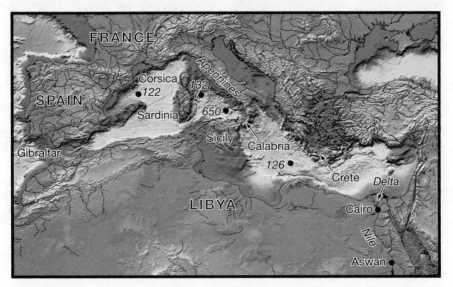

The Mediterranean sea floor, showing place-names and numbered deep-sea drilling sites relevant to this chapter and the next.

another site, whether to spend time trying to extract a very expensive string of stuck drill pipe or to abandon it and move on so they could core another site—all the time trying their best to get the very most out of their allotted two months in charge of this wonderful but very costly and temperamental geological tool.

The solution to the mystery of the M-reflector emerged gradually, and with great difficulty, over those two months. When they first reached it with the drill at site 122, off the Spanish coast, 156 meters below the sea floor in water 2,146 meters deep, the core brought up gravel with a few grains of gypsum, just as the drill got stuck—a small but important hint. Because gypsum forms by evaporation, it cannot be deposited today on the floor of the Mediterranean far below sea level.

A Wild Idea

So a debate began on board the ship: Could the sea floor once have been up at modern sea level, allowing evaporation and the deposition of the gypsum, subsequently sinking to oceanic depths? That disagreed with everything known about the behavior of the Earth's crust.

Or had the Mediterranean floor always been deep, and at some point—here was the first inkling of something new!—it got closed off from the rest of the ocean and its water evaporated away? Could the Mediterranean have become a dry desert far below sea level—a kind of Death Valley on a colossal scale? To most of the shipboard party, the idea was preposterous.

But as Leg 13 continued, they learned more and more about the M-reflector. From other holes they learned that it not only contains gypsum and halite[13] but also anhydrite—an evaporite mineral with the same composition as gypsum but with no water in the crystal structure.[14] Anhydrite is familiar as plaster of Paris. The anhydrite had a distinctive "chicken-wire" structure, known to form where anhydrite is crystallizing on the very hot shores of evaporating lagoons right at sea level. Those must have been the conditions when the M-reflector was being deposited. But was that ancient sea level the same as modern sea level, or far below it?

The answer to that question came from the sediments above and below the evaporite. It was not difficult for the *Challenger* geologists to drill down to the layers just above the M-reflector. Those layers, they found, were always marls, full of microfossils that had lived in a normal, deep-sea setting like the modern Mediterranean. Dating the microfossils showed that the change from evaporating shallow water to a deep-sea setting happened exactly at the Miocene-Pliocene boundary.[15] The evaporites must be of Late Miocene age, the same as the Messinian gypsum in the Apennines and Tuscany, and the salt in Sicily. Deepwater marls right on top of chicken-wire anhydrite looked like the record of a sudden flooding of a deep but completely dried-out Mediterranean, though it was not yet conclusive evidence.

More evidence came when they were finally able to get to the bottom of the evaporite sequence. It was not possible to drill all the way through the evaporites, for drilling progress was so slow that they simply did not have time. So Bill and Ken used a neat trick—at site 126 they drilled in a place where earlier seismic reflection surveys had a found a channel cutting down through most of the evaporite interval, so they could put down the drill bit near the base of the evaporites and drill into what lay beneath them. This showed that the sediments below the M-reflector were just like those above. They also were deep-sea marls, of a slightly older age, convincingly demonstrating that the Mediterranean sea floor was already deep when the evaporation began.

The evidence from the marls above and below the evaporites was very clear. The floor of the Mediterranean had *not* risen up to sea level and then sunk back down. It looked as if a deep Mediterranean sea floor, similar to that of today, had been exposed during a brief "salinity crisis."

As the drilling progressed, Bill Ryan, Ken Hsü, and Maria Bianca Cita, the Italian specialist in micropaleontology, gradually developed the idea that the Mediterranean had simply dried up. They knew that at the present time the Mediterranean loses more water through evaporation than is brought in by all the rivers that drain into it. The difference is made up by seawater flowing in from the Atlantic, through the Strait of Gibraltar.[16] If the strait were blocked off today by tectonic uplift, the Mediterranean would evaporate in a few centuries.

Today's Gibraltar probably did not exist in the Miocene, but there would have been some other entrance in Spain or Morocco for Atlantic water to keep the Mediterranean full. Bill, Ken, and Maria began to picture tectonic uplift closing that entrance at the beginning of the Late Miocene. Evaporation would have dried out the Mediterranean, turning it into a colossal desert about 10,000 feet below sea level. Finally, when the dam broke, a huge waterfall would have formed as the Atlantic Ocean poured in to fill the Mediterranean at the beginning of the Pliocene. Our mental picture of a waterfall barely begins to convey the idea of a colossal wall of Atlantic Ocean water surging through a gap and plunging a mile or more to the dry floor of the Mediterranean.[17]

Evaporation of the Mediterranean, followed by that catastrophic flooding, might explain everything in the cores from Leg 13. The other scientists on the ship saw various problems and were not fully convinced. In the end only Ken Hsü, Bill Ryan, and Maria Cita signed the technical articles presenting the Mediterranean deep-desiccation hypothesis.[18]

The Messinian Salinity Crisis

When *Glomar Challenger* docked at Lisbon on October 6, 1970, at the end of Leg 13, Bill, Ken, and Maria began the long struggle to test their idea of an evaporated Mediterranean and to persuade the geological world to take it seriously. To their delight, completely unexpected support soon arrived in a letter addressed to Bill Ryan.

The Leg 13 story had been reported in the *New York Times*, and it reached the attention of I. S. Chumakov in the USSR. Chumakov was the Russian geologist who had studied the wells that found the deeply incised canyon at the Aswan Dam site, now filled with silt. Chumakov wrote to tell Bill about that filled-in gorge, and to say that the greatly lowered sea level of the evaporated Mediterranean would explain how the Nile had been able to cut the strange canyon. The Gibraltar waterfall at the beginning of the Pliocene, he pointed out, would have refilled the Mediterranean and driven seawater with marine microorganisms all the way up the Nile canyon to Aswan. The mystery canyon and the mysterious marine sediments at Aswan were explained at last. The three Leg 13 scientists had solved a problem they didn't even know existed.[19]

It has been a third of a century since the voyage of *Glomar Challenger* in the Mediterranean. The Mediterranean-desert concept was one of the most dramatic geological discoveries of all time. Geologists have never stopped testing, challenging, defending, and developing it, for that is how science goes about finding out what really happened in the Earth's past. I think it would be fair to say that Mediterranean desiccation is now widely accepted, if not universally.[20]

The concept is now in a much-refined state, in which a detailed chronology supports a two-stage process.[21] An early, partial lowering of sea level produced the evaporites in places like the Apennines. A later, full sea-level fall yielded the M-reflector evaporites in the deep Mediterranean and allowed the cutting of the deep canyons on the surrounding continents.[22]

The story of the Messinian salinity crisis has two very interesting connections, leading us in both directions through the history of geological thought. First it leads us backward to Nicholas Steno, in the seventeenth century. As we saw in chapters 5 and 12, Steno's reading of the Tuscan landscape required collapse of underground caverns to make valleys, and that is now attributed by geologists to collapse of the rocks above listric extensional faults. Steno went on to infer that the sea had later flooded into the valleys.[23] We now understand that the flooding was due to the refilling of the dried-up Mediterranean at the end of the Messinian salinity crisis. Steno could not possibly have imagined the evaporation of the Mediterranean, or all the steps that have led geologists to our current understanding, but his interpretation of the Tuscan landscape was truly remarkable. Nicolaus Steno had been on the right track.

The other connection leads us forward in time, for the Messinian story inspired Bill Ryan to think carefully about other situations that might have caused seas to evaporate. He and his Lamont colleague Walter Pitman realized that the Black Sea would have become isolated as sea level fell during the ice age because of the storage of great amounts of seawater in glaciers. They found that evaporation would lower the surface of the Black Sea, which would suddenly be flooded when sea level rose again at the end of the ice age. In contrast to the evaporation of the Mediterranean, which took place about 6 million years ago, before there were any human beings to witness it, the flooding of the Black Sea was only 7,600 years ago, well within human history. Bill and Walter have made a strong case that this flooding was the source of the story of Noah's flood, and their account of it makes fascinating reading.[24]

So now, with an understanding of the coupled fronts of compression and extension sweeping across Italy, generating the folds and then tearing them apart, and with the Messinian salinity crisis punctuating Apennine history, we are in a position to ask the final question: Why did all this happen? What forces within the Earth are responsible for the building of the Apennine Mountains?

14

BEYOND PLATE TECTONICS

THE APENNINES always mystified me. They made no sense whatsoever in terms of plate tectonics. In fact nothing about Italy or the seas and islands that surround it seemed to fit very well into a plate-tectonic framework.

Plate tectonics is a set of ideas about a dynamic Earth, with continents moving around, opening new ocean basins, closing old ones, and pushing up mountain ranges. It completely revolutionized geology, and the revolution brought an exhilarating sense of excitement and intellectual adventure to our science.

Yet, plate tectonics deals with *big* features. It sees the Earth's surface as divided into a dozen or so large plates, like thin caps fitting the spherical shape of the globe. Each plate slides around, moving relative to the adjacent ones. The plates are essentially rigid and do not deform very much internally, so all the major deformations take place at plate boundaries where two plates diverge, converge, or slide past each other. Mountain ranges form at converging boundaries, where one plate is subducting under another and sinking down into the mantle. Since the boundaries between large plates extend for long distances, most mountain belts are large-scale features—either straight or gently curving.

In the Mediterranean, the maps show lots of tightly curving little mountain ranges, wrapping around intricate little oceans and bits of continent. It was very hard to attribute all this complexity and fine detail to the actions of the large plates that were explaining so much else about global geology. At first it made me wonder if the theory of plate tectonics was really valid, but as the evidence supporting the theory became overwhelming,[1] I gradually realized that the Mediterranean must be telling us about something more—about something *beyond* plate tectonics.

Microplates

Geologists like to study maps in great detail, for each map is a cornucopia of information, a treasure chest of data. We come to know maps intimately, returning again and again, looking for clues and connections and ideas. Early on, as my interest in the geology of Italy deepened and I pored over

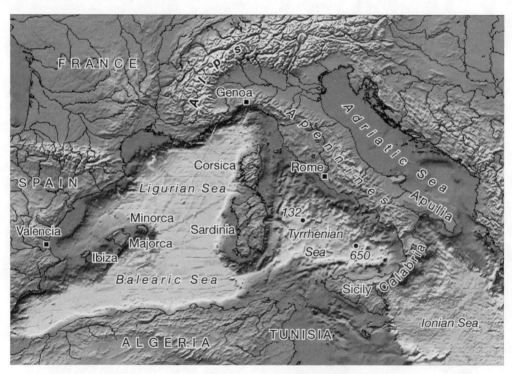

The western Mediterranean is a complex mosaic of small ocean basins separated by microcontinents like Ibiza-Majorca-Minorca and Corsica-Sardinia. Winding through the mosaic are the tightly curving mountain ranges—the Alps and the Apennines, with their continuation through Calabria into Sicily and northern Tunisia and Algeria.

maps of the Mediterranean, it struck me that the islands of Corsica and Sardinia look very strange, sitting out in the middle of the western half of the Mediterranean, separating the Ligurian and Tyrrhenian seas. Had they always been there, or had they somehow *moved* into that location? Under the old idea of fixed continents, such an idea would never have arisen, but the new plate tectonic ideas made it an obvious question.

Staring hard at the map, I began to see a pattern, something like a pair of swinging doors opened partway out from Europe. Corsica and Sardinia made one door, with a hinge at Genoa. The Balearic Islands of Spain— Majorca, Minorca, and Ibiza—made the other, with its hinge near Valencia, and not opened as far.

I knew that Corsica and Sardinia are made of continental crust, with lots of granite and old metamorphic rocks, and are quite similar to nearby parts of Europe. The Ligurian Sea, which separates them from France, was presumably made of oceanic crust. It reminded me of the new plate tectonic maps, showing the continents of North America and Europe moving apart, with the Atlantic Ocean growing in between them. But it seemed like plate tectonics in miniature, for the Corsica-Sardinia block is tiny on the scale of continents—it was what geologists were beginning to call a "microcontinent."

The swinging-door microcontinent of Corsica and Sardinia looks as if it has rotated about forty-five degrees relative to France, opening up the triangular Ligurian Sea. That turned out to be the key to testing the idea, as I learned when I tracked down earlier work on this subject. Several paleomagnetists had thought of the idea before I did, and they had already shown that the fossil magnetic compasses in volcanic rocks from Sardinia point northwest.[2] That direction is different from the north-pointing directions from France—just as you would expect if the islands were a "microplate" that had rotated relative to Europe.[3]

It seemed that microplates might be the key to understanding the Mediterranean, and I soon met a couple of Italian geologists, Tommaso Cocozza and Forese Wezel, who were thinking along the same lines. As we studied maps of the geology of the Mediterranean, we thought we could see yet another microplate. It was Calabria—the toe of the Italian boot, along with a related area in the northeast corner of Sicily.

Calabria had always puzzled Italian geologists. Geographically it forms a curving link between the Apennines, which run the length of the peninsula, and the similar mountains along the northern part of Sicily. Yet Calabria is nothing at all like the Apennines and the Sicilian ranges, which are made almost entirely of sedimentary rocks that have never been buried very deeply. Calabria, by contrast, is mostly made of granite and metamorphic rocks, much like Corsica and Sardinia.[4]

What was this chunk of continental crust doing in the sedimentary Apennines? The obvious answer would be that this was the original basement upon which the Apennine sedimentary rocks had been deposited, now uplifted and exposed to view. That was indeed the conclusion of the first geologist to make a major study of Calabria.[5] But later, more detailed examinations showed that it was the other way around. The continental crust of Calabria had been thrust on top of the sedimentary Apennines.[6] But where had Calabria originally been located, before the thrusting?

Surprisingly it was the Alps that provided the critical clue. Today the Alps curve around and terminate at the south coast of France, but in northeastern Corsica there is a little mountain belt with the same age and character as the Alps. When we moved the map of Corsica and Sardinia up against France, the Alps clearly continued into Corsica. Tommaso, Forese, and I also knew that in Calabria a little mountain-belt fragment of Alpine age had been discovered.[7] When we moved Calabria northwestward so it touched Sardinia, closing up the whole Tyrrhenian Sea, we had a restoration in which the Alpine belt was satisfyingly continuous, rather than all fragmented and scattered around.[8]

It really looks as if a little Calabrian microplate has moved southeastward about five hundred kilometers (or three hundred miles), opening up the Tyrrhenian Sea in its wake. And it even appears that Calabria is *still* moving southeastward, because there are deep earthquakes and active volcanoes under the southeastern part of the Tyrrhenian Sea—the signatures of an active subduction zone where plate movement is currently going on.

It seemed to make sense, but there was a serious problem that troubled us, and that problem came from the evaporite deposits of the Messinian salinity crisis.

At site 132, in the northwestern Tyrrhenian Sea, the Leg 13 scientists had drilled into Messinian gypsum, overlain by Pliocene marls.[9] They had not had time to drill a hole in the southeastern Tyrrhenian Sea but had tentatively concluded that Messinian evaporites cover the entire floor of the Tyrrhenian Sea. If that were true, it would mean that the whole sea has been there for at least five million years, and could not possibly be opening up behind a Calabria still in motion. If so, our Calabria story would be wrong.

In science you often walk a fine line between theory and observation. No theory can survive definitive evidence to the contrary, but what if the contrary evidence is shaky? Tommaso, Forese, and I chose to publish our idea that Calabria is a microcontinent moving southeast away from Corsica and Sardinia, and to discuss the possible conflicting evidence.[10] In that 1974 paper we made it clear if Messinian evaporites were to be found in the southeastern Tyrrhenian Sea, it would prove our hypothesis wrong, but we suggested that the southeastern part of the sea might actually be younger, postdating the Messinian, which would strongly support our hypothesis. Clearly the important test would be to drill the southeastern Tyrrhenian Sea in search of Messinian evaporites.

It took more than a decade to find out, but in 1986 the new drilling ship *Resolution* sailed to the Tyrrhenian Sea.[11] Hole 650, in the southeastern part of the Tyrrhenian Sea, found Late Pliocene resting directly on basalt of the oceanic crust. The Messinian evaporites were absent, just as we had predicted. The southeastern part of the Tyrrhenian Sea *is indeed* younger than the northwestern part.[12] It gave us a strong confirmation that Calabria has moved southeastward, opening up the Tyrrhenian Sea in its wake, and is still doing so.

Falling into the Mantle

It was very satisfying to put the Calabria piece of the puzzle in place. We did not realize at the time that we had in our hands a major clue to understanding the Apennines as well, because the Apennine coupled-front paper of Livio Trevisan and his friends did not come out until the following year.[13]

When you look at Calabria moving southeastward, you realize that in front, Calabria is compressing the sediment and crust of the Ionian Sea.[14] Behind, it is opening up the Tyrrhenian Sea in extension. Calabria is trapped between moving fronts of compression and extension, just like the Apennines. Calabria is actually much more dramatic, for it has moved hundreds of miles.

Let us first see why this kind of moving block, bounded by compression in front and extension in the rear, is so puzzling. Then we can see the kind of ideas geologists are currently thinking about as a possible solution to the puzzle.

The problem is that it is very hard to imagine what is driving the moving block. To see why, it is helpful to try a "thought experiment"— simply thinking about what would happen in an experiment so obvious that we do not actually have to set it up and try it. First, in your imagination, moisten some toilet paper and lay it on a tabletop, patting it down so it sticks loosely to the table. Then set a block of wood on the wet paper and push it to one side. What happens?

Clearly the paper will wrinkle up in front of the moving block and tear apart behind it. There will be moving, coupled fronts of compression and extension, just as in the Apennines and Calabria, because you are pushing on the block with your hand.

But in the real case of the Apennines and Calabria, there is obviously no great hand from the sky pushing the mountains outward. The forces must be coming from within the Earth, beneath the moving block. Something must be happening down there, in the mantle, to move a little block of crust sideways.

What might explain this strange behavior? The mystery has not yet been fully solved, but after the landmark 1975 Pisan paper presented the concept of the paired fronts,[15] a number of geologists came to roughly the same concept about what motions at depth could be driving the Apennine block outward.[16] The only reasonable explanation, they concluded, would be if something were falling into the mantle.

To see how it works, let's do another thought experiment. Think about slowly lowering a flat tile into a bathtub of water. As the tile descends, the water underneath it has to flow around it and fill in above. The same thing would happen in the Earth where the mantle, which can deform very slowly on geological timescales, will flow around a sinking object.

Now try the thought experiment again, but this time hold one edge of the tile up at the surface of the water to make a kind of hinge, so the tile tilts as the other side sinks. In this case the water will have to flow around the sinking end in order to fill in. If you have a block of wood floating over the tilting tile, it will be carried toward the hinge. If you had laid a paper napkin on the water surface, under the floating block, it would be compressed in front and extended behind, just as in the Apennines and Calabria.

But in the *real* Earth, what is sinking? In the case of Calabria, it would be the oceanic crust of the Ionian Sea—very much older than the Tyrrhenian crust—that is falling into the mantle. That is not particularly surprising, because one of the central features of plate tectonics is that oceanic crust does subduct—it falls into the mantle. And in fact there are deep earthquakes under the southeastern Tyrrhenian Sea showing us that there is indeed old ocean crust down there—formerly part of the Ionian Sea—sinking down into the mantle.[17]

Now we encounter a serious problem. In the Apennine case there is no oceanic crust to subduct. The Apennines lie on continental crust, and if *that* is sinking into the mantle, it is very strange, because another of the central features of plate tectonics is that continental crust floats on the mantle—it does *not* subduct.

The concept that looks best right now is that the *upper* part of the continental crust does indeed float, and is analogous to the block of wood in the second thought experiment, but the *lower* part of the continental crust peels off and sinks like the tile. This hypothetical process has come to be called "delamination."

Why would any part of continental crust sink into the mantle? In order to do so, the sinking object must have a density greater than that of the

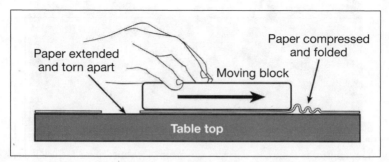

A thought experiment to show how pushing a block of wood over moist paper will compress it in front and tear it apart behind. This pattern is reminiscent of the coupled fronts of compression and extension in the Apennines and Calabria.

mantle, just as a dense tile sinks in water, while a light wooden block floats. Continental crust is less dense than mantle, which is the reason continents stay at the surface. Why would part of a continent begin to sink?

Beginning in the 1990s, geologists realized that the lower part of continental crust, which probably has a composition more like basalt than granite, will float on the mantle only if it remains at crustal depths, typically no more than thirty-five kilometers. But if it is dragged down below fifty kilometers, the increasing pressure will change basalt into a rock called eclogite, which is denser than mantle rock and will thus begin to sink.[18] If any portion of the basaltic lower continental crust gets deep enough for this to happen, it should pull the adjacent part of the lower crust down to where it also begins to sink, and the whole lower crust could peel off and delaminate. This peeling off has been called "rollback." Delamination and rollback are critical new concepts that take us beyond plate tectonics.

The deformation of the Italian continental crust that produced the Apennines seems to have begun with the collision between Corsica and Tuscany depicted in the top center panel of the second figure in this chapter. That collision would provide a way to get delamination of the Adriatic continental crust started. But is there any *evidence* that delamination and rollback are actually happening under the Apennines? Until recently the

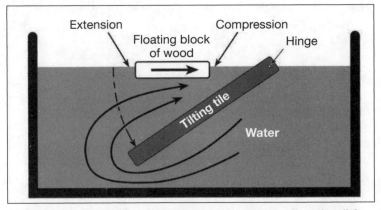

A thought experiment to show how water from beneath a sinking, tilting tile will flow around and fill in above the tile, dragging along a floating block of wood. Something like this flow pattern is probably going on in the mantle, displacing the Apennine and Calabrian blocks, so that there is compression in front of the moving block and extension behind it.

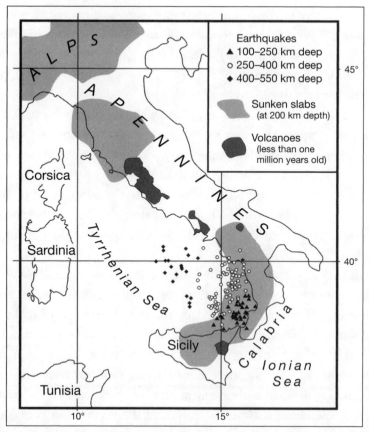

Our final map, giving the information we now have on what is going on down deep under the Apennines, Calabria, and the Tyrrhenian Sea. The sinking slabs are probably responsible for the building of the Apennines.

answer would have been no—in a place like the Apennines, which has no earthquakes at depth, there is unfortunately no way to see down into the opaqueness of the deep Earth.[19]

However, recent research has produced two ways to learn something about what is going on under the Apennines, and they both provide support for the idea of delamination. The first is a remarkable technique called seismic tomography. Tomography using X-rays has become a very important imaging tool in medicine, allowing doctors to see what is inside their patients' bodies. In a very analogous way, tomography using seismic waves from earthquakes lets us image the interior of the Earth.[20]

The Dutch seismologist Wim Spakman was the first to apply seismic tomography to the Mediterranean region, and his results were quite remarkable. Beneath the Apennines, as much as five hundred kilometers down, he could see an unusually dense, large object just where the delaminating, sinking lower crust should be.[21] It provided remarkable support for what the tectonic geologists had predicted.

Second, we have information about what is going on at depth, down where the delaminating lower crust should be, from studying volcanoes. Rome is flanked by the four large, very young volcanoes we visited early in this book, and there are several much smaller, slightly older areas of volcanic rocks and granites in Tuscany and in the islands west of Tuscany. These igneous rocks cooled from magmas that rose buoyantly from depth, and their chemistry carries lots of subtle but important clues about conditions far below the surface.

Italian geologists have studied these rocks in great detail for at least a century, but only recently have the young Italian volcanic rocks begun to make sense.[22] Teasing apart the deep sources of the igneous rocks we see at the surface, they could show that the magma came from the melting of (1) delaminated, sinking lower continental crust and upper mantle of Adria, and (2) fresh, hot, new mantle rising to fill in above the sinking Adria. It was a remarkable confirmation of the idea that delamination is driving the northeastward migration of the Apennines between their coupled fronts of compression and extension.

The Apennines have proved such a remarkable place for investigating these processes, going beyond plate tectonics, that a large international team has recently been using this part of Italy as a natural laboratory for studying the effects of delamination.[23] Their results will be published soon, and should mark the next major step forward in understanding the origin of the Mountains of Saint Francis.

The Geologist's View from a Mountaintop

Many Italian geologists now accept the concept of delamination, and this at last gives us a tentative understanding of how the Apennine Mountains were built. How far we have come!

In the Middle Ages and the Renaissance, the landscape of Italy was seen as the special handiwork of the Creator, made not so very long ago.

Or perhaps it was the wreckage of the Creator's initial handiwork, damaged by Noah's flood. For Dante in 1300, the Italian landscape roofed over a subterranean hell full of sinners, although Dante also glimpsed the water cycle that we now know erodes the mountains. In the seventeenth century Nicolaus Steno figured out how to read the history written in rocks. His successors learned to use fossils to date rocks, and found Earth history to be much longer and more complex and interesting than anyone before had realized.

The saga of understanding the Apennines has been punctuated by several revolutionary scientific breakthroughs. The discovery of thrust faults in the Alps, and later in the Apennines, pointed to a much more dynamic Earth than previously suspected. The explanation of turbidite sandstones and pelagic limestones changed the way we read Earth history recorded in rocks. The discovery of plate tectonics was the great geologic revolution, and without plate tectonics there was never any hope of understanding the Apennines. But plate tectonics alone was not enough, as the discovery of the coupled compressional and extensional faults made clear. Today we are in the middle of working out the implications of the discovery of delamination and the sinking of the lower continental crust.

I suspect there are further revolutions to come in the geological study of the Apennines, for that is the way of science. But at this point it seems as if the work of several generations of geologists has given us a real understanding of the origin of this beautiful and historic mountain range. For a geologist today, looking out from a high peak across the Mountains of Saint Francis, there are signs everywhere of Earth processes at work and vestiges everywhere of ancient worlds lost in the passage of time. It is a landscape whose beauty tells a coherent and satisfying story.

Epilogue

AUGUST 2004 was a special moment for the science of geology in Italy. Geologists from all over the world were converging on the great Medici castle of Florence for the Thirty-second International Geological Congress. Though the congress is a major scientific event, held once every four years in different places around the world, it had not met in Italy since the second congress, in Bologna in 1881. There were talks about geological research from all over the world, but the congress in Florence was largely a celebration of Italian geology, showcasing the discoveries of the Italian geologists.[1]

The progress has been dramatic and unmistakable, from the heroic pioneering days of Migliorini, Merla, Signorini, and Trevisan, which I almost overlapped, to the mature scientific world of Italian geology today. It was great to be part of the congress—hearing exciting lectures, talking to old friends, and meeting the new generation of young Italian geologists. The surprise attraction was the exhibit hall, displaying a cornucopia of detailed publications and intricate, colorful geological maps from all over Italy. With this richness of information, geologists are taking the understanding of the Apennines to still more sophisticated levels.

Shortly after the congress, Milly and I returned to Assisi, almost thirty-five years after our trip of 1970. I was writing this book and wanted to revisit the place where the story had begun.

It was the beginning of September, and late summer gave the little city of Saint Francis a completely different character. No longer dusted with snow, the streets and piazzas were thronged with people enjoying the pleasant warmth of late afternoon. From the Piazza of Santa Chiara, at the south end of town, we looked out over the plain of Santa Maria degli Angeli. Long before, from this spot, we had seen a rainbow touching down on the cathedral that encloses a little church rebuilt by the youthful Saint Francis. Just up the street, the hotel we remembered, Albergo Sole, was still there. When we recounted our winter stay so many years earlier, the innkeepers welcomed us as if we were long-lost family.

We walked through the medieval streets, past the main piazza, and on to the north end of town. Just beyond the city walls lay the great basilica—the majestic, perhaps inappropriate monument to humble Brother Francis, whose life of poverty has so deeply touched so many people.

The church was no longer frigid and empty, as on that winter day so long ago. Now it was warm and full of visitors. It was also full of the understanding we had gained over that third of a century. Giotto's frescoes told the life of a saint whose story we had come to know well. The limestone blocks reminded us of the Earth's magnetic reversals and the mass extinction they record. The ceiling rebuilt after the 1997 earthquake was a witness to the wave of compression that built the Apennine Mountains and the wave of extension that is now tearing them apart.

A road passed out through the town gate, and we walked along the back side of the hill of Assisi until we came to the old quarries that must have been the source of the limestone used to construct the Basilica of Saint Francis. Long ago we had drilled samples of *Scaglia rossa* limestone here with Bill Lowrie, as part of our paleomagnetic study. We passed a nostalgic half hour talking with Bruno Bovi, the same master stonecutter we had known in 1973, who remembered us working in his quarry so many years before.

Walking back, we reentered the town gate and came once again within sight of the Basilica of Saint Francis. Just at that moment the church was dramatically silhouetted against the brilliant orange setting sun, shining perfectly through the arches of the bell tower.

It was a sight that took our breath away, and it seemed wonderfully symbolic. It seemed to bring full circle our long Apennine journey that began at Assisi with a rainbow of promise, falling on a church that marked

the beginning of the career of Saint Francis. On that long-ago morning, when our rainbow fell on Santa Maria degli Angeli, clouds obscured the sun. Now the sky was cloudless, and the sun was clearly visible, entwined with the church that marked the culmination of the life of Saint Francis.

In those thirty-four years the geology of Italy has also gone from obscurity to clarity. In 1970 geologists had neither the local knowledge of Italy nor the general understanding of mountain building to make much sense of the Apennines. In those earlier days there was still debate over even the most basic question—whether or not the Apennine rocks have been displaced at all by thrust faults. Today geology has advanced from the exhilarating days of the plate-tectonic revolution, when we first began to sense the outline of a serious understanding of the Earth, to a mature science, rich in deep insights into the nature of our planet and its history.

During my third of a century of fascination with the Mountains of Saint Francis, the Italian geologists have come to a deep understanding of the mountains and valleys of their country. What a privilege it has been to contribute to that effort and to help them set in place a few of the stones in that edifice. And what a privilege to join in uncovering the past of our Earth, high up in the mountains with good friends, in a lovely and historic land!

NOTES

Prelude

1. To avoid unnecessary wordiness, in this book I will use "geologists" to mean all students of the Earth and its history, including paleontologists, geophysicists, geochemists, oceanographers, and atmospheric scientists.

Chapter 1. Assisi in the Winter

1. The answer to the puzzle of the flat Tiber Valley in the midst of the Apennines takes some time to develop, but it will finally emerge in chapter 12.
2. The story of those two discoveries is told in my previous book, *T. rex and the Crater of Doom* (Princeton, N.J.: Princeton University Press), 1997.
3. Alvarez, W., 2002, The greater community of scholars, extended in time, in Coons, J. E., ed., The idea of a community of scholars: Berkeley, The Faculty Club, p. 27–36.
4. The metaphor is from p. 4 of The Landscape of History (2002), Oxford University Press, by John Lewis Gaddis, whose insights into the relationship between geology and history have profoundly affected my thinking.
5. Brooke, R. B., 1959, Early Franciscan government, Elias to Bonaventure: Cambridge, Cambridge University Press, 313 p. See also Bredero, A. H., 1986, Christendom and Christianity in the Middle Ages: Grand Rapids, Michigan, William B. Eerdmans Publishing Co., p. 246–247, 251–255. The rift within the Franciscan Order that began with Brother Elias forms the background for Umberto Eco's popular medieval murder-mystery novel, The Name of the Rose (1983, San Diego, Harcourt Brace Jovanovich, 502 p.).
6. Here is the argument that the beds were originally horizontal: Wherever there is a "contact," or change from one type of sediment to another—from limestone to shale, for example—visible in a quarry or on a mountainside, that contact

must mark the surface of the sea floor when the change took place, for the new sediment type would appear on the sea floor at the same time throughout any small area. In the *Scaglia*, the contact always runs parallel to the beds in the limestone. Therefore the beds were parallel to the sea floor at the time of the change in sediment type. But the sea floor cannot have sloped more than a few degrees, or else a submarine landslide would have reduced the slope and leveled out the seabed—and indeed we find such landslides at some places in the *Scaglia*, indicating that tilting of the accumulating sediments took place but was quickly leveled out.

Chapter 2. An Invitation to Rome

1. Lanciani, R. A., 1986, La distruzione dell'antica Roma: Roma, Armando Curcio Editore, 149 p.
2. The most impressive of the archaeological maps of Rome was Rodolfo Lanciani's *Forma Urbis Romae* of 1893–1901, in forty-six sheets at the detailed scale of 1:1,000. A number of these sheets are reprinted at a reduced scale in Lanciani, R. A., 1986, La Distruzione dell'antica Roma: Roma, Armando Curcio Editore, 149 p. Some portions can now be found as scans on the Web.
3. My Berkeley colleague Bob Brentano put it this way: "The city had its seven hills. They were not always identified as the same seven hills, but the important number was seven. Numerology was more important than geography." (Brentano, R., 1974, 1990, Rome before Avignon: A social history of Thirteenth-Century Rome: Berkeley, University of California Press, p. 17.)
4. The data come from Fig. 1 in Funiciello, R., and Rosa, C., 1995, L'area romana e lo sviluppo delle ricerche geologiche: Memorie Descrittive della Carta Geologica d'Italia, v. 50, p. 23–29.
5. Ammerman, A. J., 1990, On the origins of the Forum Romanum: American Journal of Archaeology, v. 94, p. 627–645.
6. To demonstrate this you can shake a jar of muddy water and watch over the hours it takes the clay to settle out. Any disturbance of the water keeps the clay from drifting to the bottom.
7. Excavations from sites elsewhere in Italy, dating to before the founding of Rome, had shown how pottery styles changed through history. The archaeologists could thus use potsherds as a kind of fossil, to date what they found in the excavations in the Forum. This is almost exactly analogous to the way geologists use the styles of animal and plant fossils to date sedimentary rock (as discussed in the chapter about Siena).
8. Ovid, 1995, Ovid's Fasti: Roman holidays (translated with notes and introduction by Betty Rose Nagle): Bloomington, Indiana University Press, 209 p.
9. "The comparison between the monument [the wall] found by us and that described by the sources is . . . inevitable. . . . He who refuses in principle such a strong comparison, without even bothering to evaluate the archaeological data, in fact denies any other comparison between literary sources and monuments, and must take responsibility for condemning archaeology to the most complete and definitive historiographic uselessness." (Translated from Carandini, A.,

1997, La nascita di Roma: Dèi, Lari, eroi e uomini all'alba di una civiltà: Torino, G. Einaudi, p. 492–493.)

10. Of course there are both young and old basins and mountain ranges around the world. But in any local region you will find younger rocks in basins and older ones in uplifts. This explains, for example, why mines tend to be in the mountains and oil fields in the lowlands. Ores are commonly precipitated from hot waters or from molten rock deep within the Earth's crust, and are likely to be exposed in mountains where uplift and erosion have brought deep rocks to the surface. Oil comes from organic matter and is destroyed if heated to too high a temperature, so the right conditions for its preservation are in sedimentary basins that have never been buried too deeply. (For an excellent introduction to the geology of oil, see Deffeyes, K. S., 2001, Hubbert's Peak: Princeton, Princeton University Press, 208 p.)

11. Gibbon later recalled, "It was at Rome, on the fifteenth of October, 1764, as I sat musing amidst the ruins of the Capitol, while the barefooted friars were singing Vespers in the temple of Jupiter, that the idea of writing the decline and fall of the City first started to my mind" (p. 3 in Lossky, A., 1966, Introduction: Gibbon and the Enlightenment, in White, L., Jr., ed., The transformation of the Roman world. Gibbon's problem after two centuries: Berkeley, University of California Press, p. 1–29). The mention of the friars places the episode in the Aracoeli, then a Franciscan church; we know now that the Temple of Jupiter was on the Capitolium, not on the Arx.

12. Muñoz, A., 1943, L'isolamento del Colle Capitolino: Roma, Max Bretschneider, 46 p.

13. Acciaresi, P., 1911, Giuseppe Sacconi e l'opera sua massima—Cronaca dei lavori del Monumento Nazionale a Vittorio Emanuele II: Roma, Topografia dell'Unione Editrice, 314 p.

14. Pietrangeli, C., 1979, Rione X-Campitelli, Parte II: Roma, Fratelli Palombi, photograph of the Tower of Paolo III on p. 167.

15. For a good touristic guide to the geology of the city of Rome, see Heiken, G., Funiciello, R., and de Rita, D., 2005, The Seven Hills of Rome: A Geological Tour of the Eternal City: Princeton, Princeton University Press, 288 p.

16. Potter, T. W., ed., 1976, A Faliscan Town in South Etruria: Excavations at Narce, 1966–71: Rome, British School of Rome, 352 p. Tim Potter passed away far too young, a real loss to archaeology: Wallace-Hadrill, A., 2000, Timothy William Potter (6 July 1944–11 January 2000): Papers of the British School at Rome, v. 68, p. vii–xix.

Chapter 3. Witness to the Volcanic Fires of Rome

1. The word "ash," in this geological usage, refers to tiny fragments of rock erupted from a volcano, and is not to be confused with the common use of "ash" for the residue left after wood is burned.

2. Geologists always carry a hammer in the field, to break open the rock. Natural surfaces are usually weathered and perhaps overgrown with lichen and moss. The freshly broken surface is clean and lets us identify the rock. This highly

sophisticated scientific technique goes way back. In his 1824 novel *St. Ronan's Well*, Sir Walter Scott made gentle fun of the early geologists in Scotland: ". . . some rin up hill and down dale, knapping the chucky stanes to pieces wi' hammers, like sae mony road-makers run daft—they say it is to see how the warld was made!"

3. Alvarez, W., 1975, The Pleistocene volcanoes north of Rome, in Squyres, C., ed., Geology of Italy: Tripoli, Earth Science Society of the Libyan Arab Republic, p. 355–377.

4. Ross, C. S., and Smith, R. L., 1961, Ash-flow tuffs: Their origin, geologic relations, and identification: U.S. Geological Survey Professional Paper, v. 366, p. 1–81.

5. Mattias, P. P., and Ventriglia, U., 1970, La regione vulcanica dei Monti Sabatini e Cimini: Società Geologica Italiana Memorie, v. 9, p. 331–384.

6. Alvarez, W., Gordon, A., and Rashak, E. P., 1975, Eruptive source of the "Tufo rosso a scorie nere," a Pleistocene ignimbrite north of Rome: Geologica Romana, v. 14, p. 141–154.

7. Brentano, R., 1974, 1990, Rome before Avignon: A social history of Thirteenth-Century Rome: Berkeley, University of California Press, p. 95, 108.

8. This ash-flow tuff is the Tufo Giallo di Sacrofano of Mattias, P. P., and Ventriglia, U., 1970, La regione vulcanica dei Monti Sabatini e Cimini: Società Geologica Italiana Memorie, v. 9, p. 331–384.

9. Alvarez, W., 1972, The Treia Valley north of Rome: volcanic stratigraphy, topographic evolution, and geological influences on human settlement: Geologica Romana, v. 11, p. 153–176.

Chapter 4. The Quest for the Ancient Tiber River

1. This outcrop was studied geologically immediately after the work crews exposed it: De Angelis d'Ossat, G., 1946, La formazione fluvio-lacustre del Campidoglio (Roma): Bollettino dell'Ufficio Geologico d'Italia, v. 69, p. 117–127.

2. For the *Cappellaccio*, the Roman geologists use the term *Tufo del Palatino*: Marra, F. and Rosa, C., 1995, Stratigrafia e assetto geologico dell'area romana, in Funiciello, R., ed., 1995, La geologia di Roma—il centro storico (Memorie Descrittive della Carta Geologica d'Italia, v. 50): Roma, Servizio Geologico Nazionale, p. 49–118.

3. Ventriglia, U., 1971, La geologia della città di Roma: Roma, Amministrazione Provinciale di Roma, p. 31–32, 45–47; Marra, F., and Rosa, C., 1995, Stratigrafia e assetto geologico dell'area romana, in Funiciello, R., ed., 1995, La geologia di Roma—il centro storico (Memorie Descrittive della Carta Geologica d'Italia, v. 50): Roma, Servizio Geologico Nazionale, p. 89.

4. Formations can be subdivided into members and beds, or combined into groups, if that is useful. Standardized procedures for defining and naming these basic units of stratigraphy can be found in Salvador, A., 1994, International stratigraphic guide: a guide to stratigraphic classification, terminology, and procedure. 2nd ed.: Boulder, Colo., Geological Society of America, 214 p.

5. Often it is convenient to compress the horizontal scale of a cross section relative

to the vertical scale. This gives rise to "vertical exaggeration," and to slopes that appear steeper on the drawing than they really are, as is the case for this cross section of the Capitoline Hill. But that is usually a small price to pay if we can avoid a drawing that is very much wider than it is high.

6. We also found evidence for a second ancient valley hidden within the northern half of the hill. That valley must have at one time hosted a lake where volcanic pumices floated after a nearby volcanic eruption. See Alvarez, W., Ammerman, A. J., Renne, P. R., Karner, D. B., Terrenato, N., and Montanari, A., 1996, Quaternary fluvial-volcanic stratigraphy and geochronology of the Capitoline Hill in Rome: Geology, v. 24, p. 751–754.

7. More specifically the isotope potassium-40 (^{40}K) decays to argon-40 (^{40}Ar). For a full explanation of the K/Ar and ^{40}Ar/^{39}Ar methods, see Faure, G., and Mensing, T. M., 2005, Isotopes: principles and applications, 3rd ed.: Hoboken, N.J., Wiley, 897 p.

8. Funiciello, R., ed., 1995, La geologia di Roma—il centro storico (Memorie Descrittive della Carta Geologica d'Italia, v. 50, with 19 plates): Roma, Servizio Geologico Nazionale, 550 p.

9. For example, Karner, D. B., and Marra, F., 2003, ^{40}Ar/^{39}Ar dating of glacial termination V and the duration of marine isotopic stage 11, in Droxler, A. W., Poore, R. Z., and Burckle, L. H., eds., Earth's climate and orbital eccentricity; the marine isotope stage 11 question (Geophysical Monograph, v. 137): Washington, D.C., American Geophysical Union, p. 61–66; Karner, D. B., Marra, F., and Renne, P. R., 2001, The history of the Monti Sabatini and Alban Hills volcanoes; groundwork for assessing volcanic-tectonic hazards for Rome: Journal of Volcanology and Geothermal Research, v. 107, p. 185–215.

10. Meanwhile Renato Funiciello and his friends have done, for all of Rome, the kind of geological study Albert and I did for the Capitoline Hill: Heiken, G., Funiciello, R., and de Rita, D., 2005, The Seven Hills of Rome: A Geological Tour of the Eternal City: Princeton, Princeton University Press, 288 p.

11. Potassium-argon dating was largely developed at Berkeley by geologist Garniss Curtis, geophysicist Jack Evernden, and physicist John Reynolds. Dating very young rocks has long been a priority of geochronologists at Berkeley because of the great importance of determining the ages of early human fossils. Garniss Curtis, who founded the Berkeley Geochronology Center, has long focused on this goal, and some of the first Berkeley dates were done on Roman volcanic rocks, because of their very young ages: Evernden, J. F., and Curtis, G. H., 1965, The potassium-argon dating of late Cenozoic rocks in East Africa and Italy: Current Anthropology, v. 6, p. 343–385; Glen, W., 1982, The road to Jaramillo: Stanford, Stanford University Press, p. 131–137. The generosity of Ann and Gordon Getty, who are particularly interested in early hominid evolution, has made possible the level of excellence that characterizes the Berkeley Geochronology Center under Paul Renne's leadership.

12. Broecker, W. S., 1995, The glacial world according to Wally: Palisades, N.Y., Eldigio Press, 312 p.

13. Muller, R. A., and MacDonald, G. J., 2000, Ice ages and astronomical causes: New York, Springer, 318 p.

14. For more detail, see Alvarez, W., 1972, The Treia Valley north of Rome: volcanic stratigraphy, topographic evolution, and geological influences on human settlement: Geologica Romana, v. 11, p. 153–176; Alvarez, W., 1973, Ancient course of the Tiber River near Rome: an introduction to the middle Pleistocene volcanic stratigraphy of central Italy: Geological Society of America Bulletin, v. 84, p. 749–758; and Alvarez, W., 1975, The Pleistocene volcanoes north of Rome, in Squyres, C., ed., Geology of Italy: Tripoli, Earth Science Society of the Libyan Arab Republic, p. 355–377.

15. Hamblin, W. K., 1990, Late Cenozoic lava dams in the western Grand Canyon, in Beus, S. S., and Morales, M., eds., Grand Canyon geology: Oxford, Oxford University Press, p. 385–433.

16. This "what if" is an example of a counterfactual—a consideration of how a change in one parameter in history might have changed the outcome (Gaddis, J. L., 2002, The landscape of history: Oxford, Oxford University Press, 192 p.). Of course we can never know what would have happened, for there are so many interconnecting variables in history, but it is a good lesson in how very dependent all of history is on a vast network of events, each of which might have had other results.

Chapter 5. Siena and the Discovery of Earth History

1. Stopani, R., 1998, La via Francigena: storia di una strada medievale: Firenze, Le Lettere, 190 p.

2. My late Berkeley colleague Alan Dundes and Italian folklorist Alessandro Falassi have dissected the traditions of the Palio in fascinating detail: Dundes, A., and Falassi, A., 1975, La terra in piazza: An interpretation of the Palio of Siena: Berkeley, University of California Press, 265 p. In each Palio ten of the city's seventeen neighborhood contrade compete in a contest marked by medieval pageantry and by bitter, centuries-old rivalries, in a tradition that totally absorbs the Sienese. The buildup to the Palio features temporary alliances among contrade, bribery of the jockeys, and the blessing of each horse in the church of its contrada. On Palio day the greatest success for a contrada is for its horse to reach the finish line first, with or without its jockey, and thus to win the ceremonial banner also called the Palio. The next greatest reward is for its traditional enemy contrada to finish second, for coming in second is the ultimate humiliation. The traditions of the Palio are endless and complex, extending back for more than seven centuries, and they both unite and divide the people of the city. In the opening of his great novel of World War II, Herman Wouk used the Palio as a metaphor for the internecine hostilities that divided Europe on the eve of that conflict (Wouk, H., 1973, The Winds of War: Boston, Little, Brown, 885 p.).

3. Until a century ago the Via Francigena was the preferred road to Rome, despite its inconvenient route through the Tuscan hills. There are two easier routes, one along the interior valley of Valdichiana, followed today by the high-speed rail line and the Autostrada—the main highway from Florence to Rome—and the other along the coastal plain called Maremma, but both of the easier routes were long infested by malaria. Dante used these malarial regions as a metaphor

for one of the inner regions of Hell: "What sorrow there would be, if from the hospitals of Valdichiana, between July and September, and from Maremma and Sardinia, the sick were all in one pit together" (Transl. from Dante Alighieri, Inferno, Canto 29,1. 46–49). Maremma has long been a fine agricultural region, but until the eradication of malaria in the twentieth century it was a dangerously unhealthy place to seek work. In an old song of Tuscany, a young woman laments: "Everyone says: 'Maremma! Maremma!' But it seems a bitter Maremma to me. My heart always trembles when you go there, For I fear you will never come back." The crippling role played by malaria in Italy until after World War II is often overlooked today.

4. Arisi Rota, F., Brondi, A., Dessau, G., Franzini, M., Monte Amiata S.m.p.A., Stabilimento Minerario del Siele S.p.A., Stea, B., and Vighi, L., 1971, La Toscana meridionale, i giacimenti minerari: Società Italiana di Mineralogia e Petrologia, v. 27—Fascicolo Speziale, p. 357–544, esp. p. 501–503.

5. The analysis of biblical chronology is usually associated with the seventeenth-century Irish archbishop James Ussher but actually goes back to the time of Constantine (Repcheck, J., 2003, The man who found time: Cambridge, Mass., Perseus, 247 p., ch. 2).

6. Cutler, A., 2003, The seashell on the mountaintop: a story of science, sainthood, and the humble genius who discovered a new history of the earth: New York, Dutton, 228 p., is a very readable and informative biography of Nicolaus Steno.

7. These problems are treated in detail by Cutler, op. cit., especially chs. 1 and 5.

8. Cutler, op. cit., p. 60.

9. Scherz, G., 1971, Niels Stensens Reisen, in Scherz, G., ed., Dissertations on Steno as geologist: Odense, Denmark, Odense Universitetsforlag, map on p. 137.

10. Leonardo da Vinci had previously made interesting observations and drawings of rocks, although he did not reach the critical conclusions Steno did, and Leonardo's work had little effect on subsequent science because he was secretive to the point of recording his observations in mirror writing.

11. Cutler, op. cit., p. 65–66.

12. Steno wrote, "If a solid body is enclosed on all sides by another solid body, the first of the two to harden was that one which, when both touch, transferred its own surface characteristics to the surface of the other." This passage is quoted by Cutler, op. cit., p. 108, who explains clearly and compellingly how Steno reached his remarkable conclusions.

13. The second and third points are Steno's laws of "original horizontality" and "lateral continuity" (Cutler, op. cit., p. 111–112).

14. I have used exactly this approach to understanding the evolution of complicated Mediterranean tectonics, by preparing very simplified diagrams that are models reminiscent of Steno's squared-off drawings: Alvarez, W., 1991, Tectonic evolution of the Corsica-Apennines-Alps region studied by the method of successive approximations: Tectonics, v. 10, p. 936–947.

15. Of course there are many caverns where groundwater has dissolved away limestones below the Earth's surface, and sometimes their roofs collapse to form sinkholes, but these caverns never get as large as the ones envisioned by Steno.

And sometimes a volcanic eruption removes magma, and the overlying rock immediately collapses to form a caldera, like the four circular depressions of the Roman Volcanic Province. But modern geology sees no evidence for open cavities as large as the ones Steno drew in his cross sections.

16. Steno, N., 1669, op. cit.

17. Rodolico, F., 1971, Niels Stensen, founder of the geology of Tuscany, in Scherz, G., ed., Dissertations on Steno as geologist: Odense, Denmark, Odense Universitetsforlag, p. 237–243.

18. The sediment-filled valleys bring us back to the malaria mentioned in note 3 of this chapter, and to the role it played in the history of Siena. The Valdichiana (first figure in this chapter) is one of these valleys, and as a down-dropped basin, it not only trapped sediment, but also held water that could not easily drain away. The resulting swamps provided a prime habitat for the malaria-carrying anopheles mosquitoes. Coastal Maremma was malarial for a different reason. The rapid rise of sea level caused by melting of the last ice sheets about ten thousand years ago drowned many coastal valleys, which are swampy because they have not yet filled completely with sediment. The hills between Maremma and the Valdichiana are well drained by rivers and streams, so there were few habitats for anopheles mosquitoes. Thus the geological evolution of Tuscany strongly influenced which areas would become malarial and which would be healthier, and this in turn allowed the rise of a rich little city like Siena along the malaria-free route through the hills.

19. Steno, N., 1669, op. cit.

20. This trip in March 2005 offered the opportunity for me to give an honorary lecture at the University of Siena, telling the story of geological research in the Apennines in a broad humanistic context, for an audience of academics of all fields (Alvarez, W., 2005, Verso una sintesi della storia dell'uomo e della terra— Toward a synthesis of human history and geologic history: Siena, Dipartimento di Scienze della Terra, Università degli Studi di Siena, 82 p.). I am very grateful to Silvano Focardi, Francesco Antonio Decandia, Antonio Lazzarotto, Luigi Carmignani, Isabella Memmi Turbanti, Claudio Ghezzo, Armando Costantini, Fabio Sandrelli, Giuseppe Sabatini, Roberto Mazzei, and Enrico Tavarnelli for making it possible.

21. Vaccari, E., 2000, The museum and the academy: Geology and paleontology in the Accademia dei Fisiocritici of Siena during the 18th Century, in Ghiselin, M. T., and Leviton, A. E., eds., Cultures and institutions of natural history: Essays in the history and philosophy of science: San Francisco, California Academy of Sciences, p. 5–25.

22. The Accademia dei Fisiocritici (www.accademiafisiocritici.it) bears an interesting similarity to the California Academy of Sciences, founded in San Francisco more than 150 years later, for both were the creations of small, remote cities made rich by mining—the mineral wealth coming from the Colline Metallifere in the case of Siena and the gold rush in that of San Francisco—where there were people with an intense interest in understanding nature.

23. These very fossils contributed to the intellectual development of Charles Lyell, an English lawyer and geologist who passed through Siena in 1828–1829. (Lyell's

Italian trip is described by Wilson, L. G., 1972, Charles Lyell, the years to 1841: the revolution in geology: New Haven, Yale University Press, 553 p., ch. 8.) Lyell visited Siena during an extensive Italian journey that provided much of the evidence that went into his great book, the *Principles of Geology* (Lyell, C., 1830–1833, Principles of geology, three volumes, first edition [reprint 1990–1991 of the original, published in London by J. Murray]: Chicago, University of Chicago Press). Sadly Lyell never met Ambrogio Soldani, who passed away twenty years before his visit. Geologists in the English-speaking world usually honor Charles Lyell as one of the founders of geology, for his insistence that the rock record of Earth history be interpreted in terms of processes acting today. But the *Principles of Geology* is a far more complex and idiosyncratic book than that, full of wonderful observations Lyell made in Italy but reaching a conclusion that geologists today would find preposterous: Lyell believed that Earth history has oscillated through literally endless cycles, with no long-term directional trends, and that no catastrophic events ever happened in the Earth's past. Lyell argued this strange view with a lawyer's skill, as documented by Rudwick, M. J. S., 1970, The strategy of Lyell's *Principles of Geology*: Isis, v. 61, p. 4–33; Gould, S. J., 1987, Time's Arrow, Time's Cycle: Cambridge, Mass., Harvard University Press, 222 p.; and Rudwick, M. J. S., 1998, Lyell and the *Principles of Geology*, in Blundell, D. J., and Scott, A. C., eds., Lyell: the past is the key to the present: London, Geological Society (Special Publication no. 143), p. 3–15. Lyell's abhorrence of arrowlike trends in history was forgotten after even Lyell, in his later years, had to accept Darwin's model of a nonrepeating biological arrow of evolution. Nevertheless, as late as the 1980s, Lyell's influence still crippled the study of dramatic events in the Earth's past—events accepted now as crucial parts of our understanding of the evolution of the planet. One such dramatic event—the impact of a large comet or asteroid in Mexico that caused the great extinction or animals and plants 65 million years ago—first came to light through research in Italy, as we will see at Gubbio in the next chapter.

24. As we will see later, most of the older Italian rocks lack visible fossils because they were deposited far below sea level, where there is no sunlight to support a flourishing biota.

25. Steno wrote (op. cit., 1669, p. 263–4), "That there was a watery fluid, however, at a time when animals and plants were not yet to be found, and that the fluid covered all things, is proved by the strata of the higher mountains, free from all heterogeneous material [that is, fossils]."

26. Rodolico, F., 1945, La Toscana descritta dai naturalisti del Settecento: Firenze, F. Le Monnier, 351 p.

27. Soldani was actually the *co*inventor of micropaleontology, along with Jacopo Bartolomeo Beccari in Bologna, home of the oldest university in Italy—a vibrant center of geological research for the last four centuries: Vai, G. B., and Cavazza, W., eds., 2003, Four centuries of the word Geology: Ulisse Aldrovandri 1603 in Bologna: Bologna, Minerva, 327 p.

28. Soldani, A. B., 1789–1798, Testaceographiae ac zoophytographiae parvae et microscopicae (2 vols.): Siena, F. Rossi. These magnificent volumes are preserved in the library of the Accademia dei Fisiocritici in Siena.

29. Soldani, A. B., 1794, Sopra una pioggetta di sassi accaduta nella sera de' 16 giugno del MDCCXCIV in Lucignan d'Asso nel Sanese: Siena, F. Rossi, 288 p.

Chapter 6. Gubbio and the Chronology of the Past

1. Limestone is a sedimentary rock composed of the mineral calcite ($CaCO_3$), mostly produced by marine plants and animals as they build their shells. Chert is a different sedimentary rock, made of silica (SiO_2), which was produced by marine animals and plants, but which has usually dissolved and reprecipitated, sometimes as nodules—blobs of silica within limestone beds.

2. Bonarelli, G., 1891, Il territorio di Gubbio. Notizie geologiche: Roma, Tipografia Economica, 38 p.

3. Bonarelli, 1891, op. cit. p. 26, 47.

4. The story of Bonarelli's life is told by Lippi Boncambi, C., 1967, Brevi cenni sulla vita di Guido Bonarelli e le sue opere, in Lippi Boncambi, C., Signorini, R., Giovagnotti, C., Alimenti, C., and Alimenti, M., eds., Descrizione geologica dell'Umbria centrale, di Guido Bonarelli: Foligno, Poligrafica F. Salvati, p. 11–13.

5. This insight comes from my Berkeley colleague George Brimhall, who, like Bonarelli, has devoted much of his career to economic geology.

6. Bonarelli, G., 1901, Descrizione geologica dell'Umbria centrale (Opera postuma curata per incarico del Centro umbro di studi per le risorse energetiche da: C. Lippi Boncambi, R. Signorini, C. Giovagnotti, C. Alimenti, M. Alimenti, 1967): Foligno (Italy), Poligrafica F. Salvati, 156 p. Although written in 1901, this study was not published until sixteen years after Bonarelli's death in 1951.

7. Cecca, F., Cresta, S., Pallini, G., and Santantonio, M., 1990, Il Giurassico di Monte Nerone (Appennino marchigiano, Italia centrale): biostratigrafia, litostratigrafia, ed evoluzione paleogeografica, in Pallini, G., Cecca, F., Cresta, S., and Santantonio, M., eds., Atti del secondo convegno internazionale—Fossili, evoluzione, ambiente: Pergola (Italy), Comitato Centenario Raffaele Piccinini, p. 63–139.

8. We now understand why large fossils are missing in these formations. Bottom-dwelling animals like corals, clams, and snails could not live because the bottom was so deep that there was no sunlight to support a food chain based on photosynthesis. Ammonites, which sink down from the surface waters, are made of aragonite, a form of $CaCO_3$ that dissolves in deep cold waters. Only $CaCO_3$ of the stable form, calcite, could accumulate at those depths.

9. In practice the field of paleomagnetism is complicated and subtle: see chap. 5 in Lowrie, W., 1997, Fundamentals of geophysics: New York, Cambridge University Press, 354 p. (2nd edition in press); Butler, R. F., 1992, Paleomagnetism: magnetic domains to geologic terranes: Boston, Blackwell Scientific Publications, 319 p.; McElhinny, M. W., and McFadden, P. L., 2000, Paleomagnetism: continents and oceans: San Diego, Calif., Academic Press, 386 p.

10. Nairn, A. E. M., and Westphal, M., 1968, Possible implications of the palaeomagnetic study of Late Paleozoic igneous rocks of northwestern Corsica: Palaeogeography, Palaeoclimatology, Palaeoecology, v. 5, p. 179–204; De Jong,

K. A., Manzoni, M., and Zijderveld, J. D. A., 1969, Palaeomagnetism of the Alghero trachyandesites: Nature, v. 224, p. 67–69; Zijderveld, J. D. A., De Jong, K. A., and van der Voo, R., 1970, Rotation of Sardinia: Palaeomagnetic evidence from Permian rocks: Nature, v. 226, p. 933–934; Alvarez, W., 1972, Rotation of the Corsica-Sardinia microplate: Nature Physical Science, v. 235, p. 103–105.

11.· Alvarez, W., and Lowrie, W., 1974, Rotation of the Italian peninsula: Nature, v. 251, p. 285–288; ———, 1975, Paleomagnetic evidence for rotation of the Italian Peninsula: Journal of Geophysical Research, v. 80, p. 1579–1592; Channell, J. E. T., Lowrie, W., Medizza, F., and Alvarez, W., 1978, Paleomagnetism and tectonics in Umbria, Italy: Earth and Planetary Science Letters, v. 39, p. 199–210.

12. Actually the paleomagnetic vectors that I am calling fossil compasses point either northwest and down, or southeast and up, in exactly opposite directions.

13. Glen, W., 1982, The road to Jaramillo: Stanford, Stanford University Press, 459 p.

14. At the time Milly was working for Walter Pitman, the leader of the Lamont magnetics group, and she drew the first world map of the magnetic stripes on the ocean floor: Pitman, W. C., Larson, R. L., and Herron, E. M., 1974, The age of the ocean basins (Geological Society of America Map and chart series MC-6): Geological Society of America, Boulder, Colo., cartography by M. M. Alvarez and H. Cason. Because of that epochal map, there was a while when Milly was much better known in the geological world than I was.

15. This account is based on Renz's own introduction to his PhD thesis (Renz, O., 1951, Ricerche stratigrafiche e micropaleontologiche sulla Scaglia (Cretaceo Superiore-Terziario) dell'Appennino centrale [Italian translation of the thesis in German, 1936]: Memorie Descrittive della Carta Geologica d'Italia, v. 29, p. 11–12), and on an obituary in Spanish written in Venezuela, where Renz worked for many years (www.pdvsa.com/lexico/pioneros/renz.htm).

16. Renz, O., 1951, op. cit., p. 12.

17. Luterbacher, H. P., and Premoli Silva, I., 1962, Note préliminaire sur une revision du profil de Gubbio, Italie: Rivista Italiana di Paleontologia e Stratigrafia, v. 68, p. 253–288.

18. Premoli Silva, I., and Sliter, W. V., 1995, Cretaceous planktonic foraminiferal biostratigraphy and evolutionary trends from the Bottaccione Section, Gubbio, Italy: Palaeontographia Italica, v. 82, p. 1–89.

19. Arthur, M. A., and Fischer, A. G., 1977, Upper Cretaceous-Paleocene magnetic stratigraphy at Gubbio, Italy. I. Lithostratigraphy and sedimentology: Geological Society of America Bulletin, v. 88, p. 367–371; Premoli Silva, I., 1977, II. Biostratigraphy: op. cit., p. 371–374; Lowrie, W., and Alvarez, W., 1977, III. Upper Cretaceous magnetic stratigraphy: op. cit., p. 374–377; Roggenthen, W. M., and Napoleone, G., 1977, IV. Upper Maastrichtian-Paleocene magnetic stratigraphy: op. cit., p. 378–382; Alvarez, W., Arthur, M. A., Fischer, A. G., Lowrie, W., Napoleone, G., Premoli Silva, I., and Roggenthen, W. M., 1977, V. Type section for the Late Cretaceous-Paleocene geomagnetic reversal timescale: op. cit., p. 383–389.

20. Pialli, G. P., ed., 1976, Paleomagnetic stratigraphy of pelagic carbonate sediments, Società Geologica Italiana Memorie, v. 15, 128 p.

21. Luterbacher, H. P., and Premoli Silva, I., 1964, Biostratigrafia del limite cretaceo-

terziario nell'Appennino centrale: Rivista Italiana di Paleontologia e Stratigrafia, v. 70, p. 67–128.

22. *K* is the standard abbreviation for the Cretaceous (*Kreide*, or chalk, in German). *T* stands for Tertiary. Recently the Tertiary, comprising the Paleocene to the Pliocene, has been formally discarded in favor of the Paleogene, including the Paleocene, Eocene, and Oligocene. However, many geologists are very used to the term "KT boundary," so I continue to use it, instead of the now-correct "KP boundary," with *P* for Paleogene.

23. Ammonites and dinosaurs left no fossils in the *Scaglia* limestone, but we knew that their extinctions were very nearly—exactly, in all likelihood—the same age as the foram extinction recorded at Gubbio, based on two centuries of construction of the geological timescale by generations of geologists. It is now clear that today's birds are descended from a branch of the dinosaurs, so careful paleontologists sometimes speak of the extinction of the "nonavian dinosaurs."

24. The extreme preference of geologists for slow, gradual Earth processes was due to the influence of Charles Lyell, as discussed in note 23 of chapter 5. This aspect of the Lyellian heritage seriously retarded understanding of the Earth for more than a century.

25. Many geologists, seeking to avoid thinking about catastrophes, argued that there was a great deal of stratigraphic section missing at the KT boundary everywhere around the world, so that the extinction only *looked* sudden. But in the *Scaglia* we found all the reversals seen in the ocean-floor record, so clearly there was little or no time unrecorded, and the extinction had to have been catastrophic.

26. Alvarez, W., 1997, T. rex and the Crater of Doom: Princeton, N.J., Princeton University Press, 185 p.

27. Terry Engelder, a friend of Bill Lowrie's and mine from Lamont, is now at Penn State. He spent two summers in Italy with me, working on the evolution of the Apennines—the topic of part 4 of this book (Alvarez, W., Engelder, T., and Lowrie, W., 1976, Formation of spaced cleavage and folds in brittle limestone by dissolution: Geology, v. 4, p. 698–701; Alvarez, W., Engelder, T., and Geiser, P. A., 1978, Classification of solution cleavage in pelagic limestones: Geology, v. 6, p. 263–266).

28. Alvarez, L. W., Alvarez, W., Asaro, F., and Michel, H. V., 1980, Extraterrestrial cause for the Cretaceous-Tertiary extinction: Science, v. 208, p. 1095–1108.

29. Smit, J., and Klaver, G., 1981, Sanidine spherules at the Cretaceous-Tertiary boundary indicate a large impact event: Nature, v. 292, p. 47–49; Montanari, A., Hay, R. L., Alvarez, W., Asaro, F., Michel, H. V., Alvarez, L. W., and Smit, J., 1983, Spheroids at the Cretaceous-Tertiary boundary are altered impact droplets of basaltic composition: Geology, v. 11, p. 668–671; Bohor, B. F., Foord, E. E., Modreski, P. J., and Triplehorn, D. M., 1984, Mineralogic evidence for an impact event at the Cretaceous-Tertiary boundary: Science, v. 224, p. 867–869. Many other kinds of evidence came to light through the efforts of many other scientists, as detailed in Alvarez, 1997, op. cit.

30. Penfield, G. T., and Camargo-Zanoguera, A., 1981, Definition of a major igneous zone in the central Yucatán platform with aeromagnetics and gravity: Society of Exploration Geophysicists Technical Program, Abstracts, and Biographies, v. 51,

p. 37; Hildebrand, A. R., Penfield, G. T., Kring, D. A., Pilkington, M., Camargo-Zanoguera, A., Jacobsen, S. B., and Boynton, W. V., 1991, Chicxulub crater: a possible Cretaceous/Tertiary boundary impact crater on the Yucatán Peninsula, Mexico: Geology, v. 19, p. 867–871.

31. As discussed in note 11 of chapter 4, potassium-argon dating was developed into a practical chronological tool at Berkeley in the mid-twentieth century by geologists Garniss Curtis and Jack Evernden, physicist John Reynolds, and paleontologist Don Savage: Glen, W., 1982, The road to Jaramillo: Stanford, Stanford University Press, 459 p. Argon is a noble gas that is not incorporated into growing mineral grains, so the minerals start out with potassium but no argon. All the argon found in an old mineral grain must have formed by radioactive decay of the potassium in the mineral.

32. In the 1980s Sandro's age-dating work concentrated on the Eocene and the Oligocene (Montanari, A., Deino, A. L., Drake, R. E., Turrin, B. D., DePaolo, D. J., Odin, G. S., Curtis, G. H., Alvarez, W., and Bice, D. M., 1988, Radio-isotopic dating of the Eocene-Oligocene boundary in the pelagic sequence of the Northern Apennines, in Premoli Silva, I., Coccioni, R., and Montanari, A., eds., The Eocene-Oligocene boundary in the Marche-Umbria Basin (Italy): Ancona, International Subcommision on Paleogene Stratigraphy of the International Union of Geological Sciences, F.lli Aniballi Publishers, p. 195–208; Premoli Silva, I., Coccioni, R., and Montanari, A., 1988, The Eocene-Oligocene boundary in the Marche-Umbria Basin, IUGS Special Publication: Ancona, F.lli Aniballi Publishers, p. 268). In the 1990s the Miocene was the interval of the most interest (Montanari, A., Deino, A., Coccioni, R., Langenheim, V. E., Capo, R., and Monechi, S., 1991, Geochronology, Sr isotope stratigraphy, magnetostratigraphy, and plankton stratigraphy across the Oligocene-Miocene boundary in the Contessa section [Gubbio, Italy]: Newsletters in Stratigraphy, v. 23, p. 151–180; Montanari, A., Odin, G. S., and Coccioni, R., 1997, Miocene Stratigraphy: An Integrated Approach, Development in Palaeontology and Stratigraphy, n° 15: Amsterdam, Elsevier, 694 p.).

33. www.geo.vu.nl/~smit/coldigioco/coldigioco.htm; www.carleton.edu/departments/geol/RelatedPrograms/coldigioco/index.html.

34. Arduino's names for the earliest part of geologic time, "Primary" and "Secondary," have long been obsolete. His third division, "Tertiary," was in use when we started working at Gubbio, but has subsequently been replaced. Although now outmoded, Arduino's timescale was a major advance in the development of geology as a science.

35. The Cenozoic Era used to be divided into the Tertiary and the Quaternary periods—the latter added to Arduino's original threefold timescale. In the latest version of the timescale (Gradstein, F., Ogg, J., Smith, A., and others, 2005, A geological time scale 2004: Cambridge, Cambridge University Press, 589 p.) the Tertiary has become obsolete and should be abandoned. Now the Cenozoic Era should be divided into the Paleogene and Neogene Periods. With apologies I keep "Tertiary" in this book because we used it so extensively in the Apennine work, and the name "Cretaceous-Tertiary boundary" (KTB) is so embedded in the geological literature.

36. Gradstein et al., op. cit.; www.stratigraphy.org; www.chronos.org; www.ucmp .berkeley.edu/help/timeform.html; and www.stratigraphy.org/geowhen.

Chapter 7. From Winter Storm to Earth Storm

1. Cecca, F., Cresta, S., Pallini, G., and Santantonio, M., 1990, Il Giurassico di Monte Nerone (Appennino marchigiano, Italia centrale): biostratigrafia, litostratigrafia, ed evoluzione paleogeografica, in Pallini, G., Cecca, F., Cresta, S., and Santantonio, M., eds., Atti del secondo convegno internazionale—Fossili, evoluzione, ambiente: Pergola (Italy), Comitato Centenario Raffaele Piccinini, p. 63–139; Alvarez, W., 1989, Evolution of the Monte Nerone seamount in the Umbria-Marche Apennines: 1. Jurassic-Tertiary stratigraphy: Società Geologica Italiana Bollettino, v. 108, p. 3–21. In addition the younger members of Professor Pallini's group, Fabrizio Cecca, Stefano Cresta, and Massimo Santantonio, have each made major contributions on their own.
2. The term *anticline* means that the beds on either side of the fold are tilted ("inclined") away ("anti") from each other. A downfold is called a *syncline* because the beds on either side are inclined toward ("syn") each other.
3. Alvarez, W., and Lowrie, W., 1984, Magnetic stratigraphy applied to synsedimentary slumps, turbidites, and basin analysis: The Scaglia limestone at Furlo (Italy): Geological Society of America Bulletin, v. 95, p. 324–336.

Chapter 8. Rocks for Building a Mountain Range

1. Lowrie, W., Alvarez, W., Premoli Silva, I., and Monechi, S., 1980, Lower Cretaceous magnetic stratigraphy in Umbrian pelagic carbonate rocks: Geophysical Journal, v. 60, p. 263–281; Stewart, K. G., and Alvarez, W., 1991, Mobile-hinge kinking in layered rocks and models: Journal of Structural Geology, v. 13, p. 243–259.
2. Geologists make a distinction between minerals and rocks. Minerals are little grains, each of which has a particular chemical composition or a limited range of chemical compositions, and a specific crystal structure. For example, quartz is a mineral with the composition SiO_2 and a distinctive crystal arrangement of the silicon and oxygen atoms. Feldspars are minerals with the composition $KAlSi_3O_8$, $NaAlSi_3O_8$, $CaAl_2Si_2O_8$, or some intermediate combination, and with closely related crystal structures. Rocks, on the other hand, are aggregates of minerals and may be sedimentary (if the mineral grains were carried to their final resting place by water, wind, or ice), metamorphic (if the original mineral grains have changed through chemical reactions), or igneous (if the mineral grains crystallized out of an original rock that was heated so hot that it melted). For example, a rock made of sand-size grains of quartz and feldspar is called granite if it is igneous, gneiss if metamorphic, or sandstone if sedimentary. Specialists in the fields of mineralogy and petrology draw much finer distinctions than in this simplification, and the fine distinctions allow geologists to understand in detail the origins and significance of very subtle variations in Earth materials. We often find that a rock has gone through different stages in its evolution—for example

a volcanic rock that later was buried, heated, and squeezed could be called a metavolcanic rock, and the terms "igneous," "metamorphic," and "sedimentary" usually refer to the most recent stage in a rock's history.

3. Kligfield, R., Carmignani, L., and Owens, W. H., 1981, Strain analysis of a Northern Apennine shear zone using deformed marble breccias: Journal of Structural Geology, v. 3, p. 421–436.

4. Cocozza, T., Jacobacci, A., Nardi, R., and Salvadori, I., 1974, Schema stratigrafico-strutturale del Massiccio Sardo-Corso e minerogenesi della Sardegna: Società Geologica Italiana Memorie, v. 13, p. 85–186; Carmignani, L., Cocozza, T., Ghezzo, C., Pertusati, P. C., and Ricci, C. A., 1982, Guida alla geologia del Paleozoico sardo: Cagliari, Società Geologica Italiana, 215 p.; Carmignani, L., Oggiano, G., Barca, S., Conti, P., Salvadori, I., Eltrudis, A., Funedda, A., and Pasci, S., 2001, Geologica della Sardegna: Memorie Descrittive della Carta Geologica d'Italia, v. 60, 283 p.

5. The best introduction to petroleum geology I know of is Deffeyes, K. S., 2001, Hubbert's Peak: Princeton, Princeton University Press, 208 p. Anticlinal upfolds may trap oil and gas, because most deep rocks have water in their pores. Oil floats on water, and gas is lighter still, so if an impermeable rock has been folded into an anticline, oil and gas may be caught in the pores of the rock layers beneath it, like air trapped in an inverted drinking glass pushed down into a bathtub full of water.

6. Martinis, B., and Pieri, M., 1964, Alcune notizie sulla formazione evaporitica del Triassico superiore nell'Italia centrale e meridionale: Società Geologica Italiana Memorie, v. 4, p. 649–678.

7. Alighieri, D., Purgatorio, Canto 14, lines 34–36.

8. The Wilson cycle was named for the Canadian geologist J. Tuzo Wilson, who, during the early days of the plate-tectonic revolution, first recognized this fundamental character of Earth history: Wilson, J. T., 1966, Did the Atlantic close and then re-open?: Nature, v. 211, p. 676–681.

9. The pellets are white and opaque because each is made of very fine-grained white mud. The composition of the mud is $CaCO_3$, corresponding to the mineral calcite. The pellets are held together by larger grains of calcite that have crystallized in the pore spaces between the pellets. Light entering the transparent coarse grains of the calcite cement can get absorbed, so the cement looks dark under the hand lens.

10. D'Argenio, B., 1970, Evoluzione geotettonica comparata tra alcune piattaforme carbonatiche dei Mediterranei Europeo ed Americano: Atti dell'Accademia Pontaniana, N. S., v. 20, p. 3–34; D'Argenio, B., De Castro, P., Emiliani, C., and Simone, L., 1975, Bahamian and Apenninic limestones of identical lithofacies and age: American Association of Petroleum Geologists Bulletin, v. 59, p. 524–530.

11. Marls are mixtures, intermediate between pure limestones and pure clays, and Fucoids are the tracks and trails, sometimes seen on these beds, left by bottom-dwelling organisms.

12. Alvarez, W., and Lowrie, W., 1984, Magnetic stratigraphy applied to synsedi-mentary slumps, turbidites, and basin analysis: The Scaglia limestone at Furlo

(Italy): Geological Society of America Bulletin, v. 95, p. 324–336; Alvarez, W., Colacicchi, R., and Montanari, A., 1985, Synsedimentary slides and bedding formation in Apennine pelagic limestones: Journal of Sedimentary Petrology, v. 55, p. 720–734.

13. Argand, E., 1916, Sur l'arc des Alpes Occidentales: Eclogae Geologicae Helvetiae, v. 14, p. 145–191; Argand, E., 1924, La Tectonique de l'Asie: Proceedings of the 13th International Geological Congress, v. 1, p. 171–372; Channell, J. E. T., and Horváth, F., 1976, The African/Adriatic Promontory as a palaeogeographic premise for Alpine orogeny and plate movements in the Carpatho-Balkan Region: Tectonophysics, v. 35, p. 71–101.

14. Channell, J. E. T., D'Argenio, B., and Horváth, F., 1979, Adria, the African Promontory, in Mesozoic Mediterranean paleogeography: Earth-Science Reviews, v. 15, p. 213–292; D'Argenio, B., Horvath, F., and Channell, J. E. T., 1980, Palaeotectonic evolution of Adria, the African promontory: Bureau de Recherches Géologiques et Minières, Mémoires, v. 115, p. 331–351.

15. A surprising feature of Adria made possible the deposition of the rare, deepwater pelagic limestones with all their precious historical record. The original continental crust of Adria was at about sea level, and thus considerably higher than the oceanic crust of the deep seas that nearly surrounded it. Rocks do not have the strength to support such differences in elevation over long periods of time, so the continental crust slowly expanded out toward the adjacent oceans. As it expanded, it got thinner, and its top surface thus gradually sank farther and farther below sea level. This happens because continental crust floats in the denser rocks of the mantle, and thinner crust floats lower, just as a thin piece of floating wood protrudes from a tank of water less than the top of a thick piece of wood. It was this gradual sinking of the top of the crust of Adria that made space for the deposition of seven hundred meters of shallow-water *Calcare Massiccio*, and then of several hundred meters more of deeper-water pelagic limestones. Bruno D'Argenio and I wrote a paper quantifying this thinning and subsidence of the Adriatic continental crust: D'Argenio, B., and Alvarez, W., 1980, Stratigraphic evidence for crustal thickness changes on the southern Tethyan margin during the Alpine cycle: Geological Society of America Bulletin, v. 91, p. 681–689; 2558–2587.

Chapter 9. Distant Thunder from the Alps

1. Vai, G. B., and Cavazza, W., 2003, Four centuries of the word Geology: Ulisse Aldrovandi 1603 in Bologna: Bologna, Minerva Edizioni, 327 p.

2. Tyler, J. E., 1930, The Alpine passes: the Middle Ages (962–1250): Oxford, B. Blackwell, 188 p.; Guichonnet, P., 1980, Histoire et civilisations des Alpes; v. 1. Destin historique: Toulouse, Privat, 417 p.; Pauli, L., 1984, The Alps: archaeology and early history: London, Thames and Hudson, 304 p.

3. There are many accounts of the German investiture controversy. I find the version of Norman Cantor to be particularly clear and informative: Cantor, N. F., 1969, Medieval history (2nd edition): New York, Macmillan, p. 293–304. For a broader view of the character of the Gregorian reform movement, see Bartlett,

R., 1993, The making of Europe: conquest, colonization and cultural change 950–1350: Princeton, Princeton University Press, 431 p.

4. Tyler, op. cit., p. 30.

5. Cantor, op. cit., p. 271–278.

6. A well-known expression in Europe, "to go to Canossa," means abjectly to beg forgiveness.

7. Ohler, N., 1989, The medieval traveller: Woodbridge, UK, Boydell Press, p. 118–123.

8. Nicolson, M. H., 1959, Mountain gloom and mountain glory: the development of the aesthetics of the infinite: Ithaca, N.Y., Cornell University Press, p. 3.

9. Nicholson, op. cit., ch. 4–6.

10. Gould, S. J., 1987, Time's Arrow, Time's Cycle: Cambridge, Mass., Harvard University Press, p. 43. Gould stresses that Thomas Burnet (1635–1715), who held this view of mountains as ruins, was as strongly committed to explaining Earth history in accordance with natural law as is any scientist today. He was forced to a catastrophic viewpoint not as a result of religious dogmatism, but because of the fact that deep time had not yet been discovered (Gould, op. cit, ch. 2). See also Rossi, P., 1984, The dark abyss of time: the history of the earth & the history of nations from Hooke to Vico: Chicago, University of Chicago Press, 338 p.; Repcheck, J., 2003, The man who found time: Cambridge, Mass., Perseus, 247 p.

11. Castellarin, A., Guy, F., and Selli, L., 1982, Geologia dei dintorni del Passo di S. Nicolò e della Valle di Contrin (Dolomiti), in Castellarin, A., and Vai, G. B., eds., Guida alla geologia del Sudalpino centro-orientale: Bologna, Società Geologica Italiana, Guide Geologiche Regionali, p. 231–242.

12. Castellarin, A., Del Monte, M., and Frascari, F., 1974, Cosmic fallout in the "hard grounds" of the Venetian region (Southern Alps): Giornale di Geologia, v. 39, p. 333–346. This paper used cosmic fallout debris as evidence for nondeposition rather than erosion, thus predating by several years the technique we used at Gubbio (Alvarez, L. W., Alvarez, W., Asaro, F., and Michel, H. V., 1980, Extraterrestrial cause for the Cretaceous-Tertiary extinction: Science, v. 208, p. 1095–1108).

13. The majestic scenery of the Dolomites has come from the erosion of younger, weaker sediments that once filled in around and over the limestone platforms. The original calcite ($CaCO_3$) of the limestone platforms has largely been replaced by the mineral dolomite ($CaMg(CO_3)_2$). Both the mineral and the mountains were named for the pioneering French geologist Déodat de Dolomieu (1750–1801).

14. Italian geologists call these deep rocks the Ivrea-Verbano zone: Salisbury, M. H., and Fountain, D. M., eds., 1988, Exposed cross sections of the continental crust: Dordrecht, Kluwer, p. 662; Quick, J. E., Sinigoi, S., Snoke, A. W., Kalakay, T. J., Mayer, A., and Peressini, G., 2003, Geologic map of the southern Ivrea-Verbano Zone, northwestern Italy, Geologic Investigations Series, U.S. Geological Survey, Report: I-2776, 22 pp., 1 sheet.

15. Wegener, A., 1912, Die Entstehung der Kontinente: Geologische Rundschau, v. 3, p. 276–292; Wegener, A., 1920, Die Entstehung der Kontinente und Ozeane: Braunschweig, F. Vieweg, 135 p. (transl. as Wegener, A., 1966, The origin of continents and oceans: New York, Dover, 246 p).

16. van Waterschoot van der Gracht, W. A. J. M., 1928, Theory of continental drift: Tulsa, American Association of Petroleum Geologists, 240 p. Wegener had proposed that his drifting continents somehow forced their way through the rocks of the underlying mantle, like ships plowing through the ocean. That was correctly rejected by geologists, but in dismissing that proposed mechanism they were led to reject a great deal of Wegener's observational evidence for continental drift, evidence that we now know is largely correct, like the fit of the continents on the two sides of the Atlantic Ocean. A few prominent geologists accepted Wegener's continental drift early on, notably Émile Argand, 1924, La Tectonique de l'Asie: Proceedings of the 13th International Geological Congress, v. 1, p. 171–372.

17. Bailey, E. B., 1935 (reprinted 1968), Tectonic essays, mainly Alpine: Oxford, Oxford University Press, 200 p. This is a delightful little classic of our science, telling the story of the early geological discoveries in the Alps.

18. Bailey, op. cit., ch. 4, 5.

19. Seismic exploration has been critical in recent advances in understanding the thrust structure of the Alps: Pfiffner, O. A., Lehner, P., Heitzmann, P., Mueller, S., and Steck, A., eds., 1997, Deep structure of the Swiss Alps: results of NRP 20: Basel, Birkhauser Verlag, 380 p.

20. Much of the early understanding of arrays of thrust faults came from the thrust belt of the Canadian Rockies, an oil-producing region where drill-hole information and seismic data were so abundant that the geometry of the thrusts and their history of emplacement could be worked out in detail (Bally, A. W., Gordy, P. L., and Stewart, G. A., 1966, Structure, seismic data, and orogenic evolution of southern Canadian Rocky Mountains: Bulletin of Canadian Petroleum Geology, v. 14, p. 337–381). Sophisticated computer techniques are now available for unraveling the structure of thrust belts: Geiser, J., Geiser, P. A., Kligfield, R., Ratliff, R., and Rowan, M., 1988, New applications of computer-based section construction; strain analysis, local balancing, and subsurface fault prediction: The Mountain Geologist, v. 25, p. 47–59.

21. Argand, 1924, op. cit.; Argand, E., 1916, Sur l'arc des Alpes Occidentales: Eclogae Geologicae Helvetiae, v. 14, p. 145–191; Schaer, J. P., 1991, Émile Argand 1879–1940: Life and portrait of an inspired geologist: Eclogae Geologicae Helvetiae, v. 84, p. 511–534.

22. Rutten, M. G., 1969, The geology of western Europe: Amsterdam, Elsevier, p. 202–204. This was the last great synthesis of Alpine geology before the plate-tectonic revolution swept away continental fixism and opened up new approaches to understanding mountain building. Another collection of papers about gravity sliding, some bearing on the Alps, is De Jong, K. A., and Scholten, R., 1973, Gravity and tectonics: New York, John Wiley and Sons, 502 p.

23. Lightman, A., and Gingerich, O., 1992, When do anomalies begin?: Science, v. 255, p. 690–695.

24. Lightman and Gingerich were using the word "theory" in the technical sense of a well-developed and strongly supported explanation for a large collection of scientific observations, not in the popular sense of a personal opinion that may have very little justification.

25. Alvarez, W., 1997, T. rex and the Crater of Doom: Princeton, N.J., Princeton University Press, ch. 3.

26. Heezen, B. C., and Tharp, M., 1961, Physiographic diagram of the South Atlantic Ocean, the Caribbean Sea, the Scotia Sea, and the eastern margin of the South Pacific Ocean: New York, Geological Society of America.

27. Hess's idea was originally presented in a widely circulated preprint: Hess, H. H., December 1960, The evolution of ocean basins, Princeton University, Department of Geology, 38 p., and was eventually published as Hess, H. H., 1962, History of ocean basins, in Engel, A. E. J., James, H. L., and Leonard, B. F., eds., Petrologic studies: a volume in honor of A. F. Buddington, Geological Society of America, p. 599–620.

28. Dietz, R. S., 1961, Continent and ocean basin evolution by the spreading of the sea floor: Nature, v. 190, p. 854–857.

29. Glen, W., 1982, The road to Jaramillo: Stanford, Stanford University Press, 459 p.; Oreskes, N., 2001, Plate tectonics: an insider's history of the modern theory of the Earth: Boulder, Colo., Westview Press, p. 424.

30. Ch. 6 in this book.

31. An alternate hypothesis would be that the Earth is expanding. This hypothesis was forcefully argued by S. W. Carey (1976, The expanding earth: Amsterdam, Elsevier, 488 p.). It has not been accepted by geologists, however, because of its physical implausibility and because subduction of old ocean crust at trenches clearly balances sea-floor spreading.

32. The old crust stays cold and rigid for a long time on its descent, and thus it is brittle and able to fracture and generate earthquakes. Plots of the positions of earthquakes show very clearly the positions at depth of the subducting slabs of old ocean crust.

33. Dewey, J. F., and Bird, J. M., 1970, Mountain belts and the new global tectonics: Journal of Geophysical Research, v. 75, p. 2625–2647; Dewey, J. F., and Horsfield, B., 1970, Plate tectonics, orogeny and continental growth: Nature, v. 225, p. 521–525.

34. Isacks, B., Oliver, J., and Sykes, L. R., 1968, Seismology and the new global tectonics: Journal of Geophysical Research, v. 73, p. 5855–5899; Barazangi, M., and Dorman, J., 1969, World seismicity maps compiled from ESSA, Coast and Geodetic Survey, epicenter data, 1961–1967: Seismological Society of America Bulletin, v. 59, p. 369–380.

35. Geologists have been able to reconstruct a great ocean we call the Tethys, now vanished, which once separated Europe and Asia from Africa, Arabia, India, and Australia. We recognize smaller oceanic areas within the Tethys, including the Pennine Ocean, and also the Liguride Ocean, which will be of importance in ch. 11. There were also microcontinents within the Tethys Ocean, a complication omitted here.

36. Maybe a sandwich is not a very good analogy to the Alps, because sandwiches are made by piling the ingredients on top of one another. To make a sandwich *really* Alpine-style, you would have to lay two slices of bread a little way apart on a table and smear jelly on the table between then. Then you would slide the pieces of bread together, through the jelly, pushing one piece of bread over the other

and squishing the jelly between them. This seems to work better in mountain ranges than in the kitchen.

37. Pitman, W. C., III, and Talwani, M., 1972, Sea-floor spreading in the North Atlantic: Geological Society of America Bulletin, v. 83, p. 619–646.

38. Dewey, J. F., Pitman, W. C., III, Ryan, W. B. F., and Bonnin, J., 1973, Plate tectonics and the evolution of the Alpine system: Geological Society of America Bulletin, v. 84, p. 3137–3180.

39. Burbank, D. W., and Anderson, R. S., 2001, Tectonic geomorphology: Oxford, Blackwell, ch. 10, esp. p. 212–224.

40. This approach, in which the entire mountain belt is seen as an "accretionary wedge," has proved extremely useful for understanding the origin of mountains (Platt, J. P., 1986, Dynamics of orogenic wedges and the uplift of high-pressure metamorphic rocks: Geological Society of America Bulletin, v. 97, p. 1037–1053).

41. Willett, S. D., Schlunegger, F., and Picotti, V., 2006, Messinian climate change and erosional destruction of the central European Alps: Geology, v. 34, p. 613–616.

42. This approach to understanding mountains is part of a fairly new branch of geology called "tectonic geomorphology," in which *geomorphology* is the study of landscapes (Burbank and Anderson, op. cit.).

Chapter 10. The Approach of Destiny

1. In the library at Berkeley, I could not find any confirmation of a connection between the Countess Matilda and the Contessa Valley at Gubbio, so I consulted two historians of Gubbio. Art historian Ettore Sannipoli, who as a boy used to help Bill Lowrie and me collect paleomagnetic samples, explored the archives of Gubbio and could find no relevant documents. Euro Puletti, an expert in the origin of place names in this part of Italy, told me that he also has never heard of any document from the time of Matilda bearing on the origin of the name. Prof. Puletti stressed that the first documented use of the name "Contessa" for this valley goes back only to the sixteenth century, half a millennium after the time of the Countess Matilda. He also pointed out that there have been numerous local countesses for whom the valley could have been named, and that "Contessa" might even be a corruption of some other word having nothing to do with a countess.

2. Here I am referring to sandstone made primarily of grains of quartz (SiO_2) and other silicon-rich minerals like feldspars. Geologists usually designate calcite-rich sands with the name "calcarenite," where "arenite" means sandstone. At some places in the Umbria-Marche limestone sequence there are sandstones in which the grains are made of calcite (Alvarez, W., and Lowrie, W., 1984, Magnetic stratigraphy applied to synsedimentary slumps, turbidites, and basin analysis: The *Scaglia* limestone at Furlo (Italy): Geological Society of America Bulletin, v. 95, p. 324–336), but quartz-feldspar sandstones are simply absent before the Miocene.

3. Bonarelli, G., 1967, Descrizione geologica dell'Umbria centrale (Opera postuma

curata per incarico del Centro umbro di studi per le risorse energetiche da: C. Lippi Boncambi, R. Signorini, C. Giovagnotti, C. Alimenti, M. Alimenti): Foligno (Italy), Poligrafica F. Salvati, 156 p.

4. Renz, O., 1936, Stratigraphische und mikropaleontologische Untersuchung der *Scaglia* (Obere Kreide-Tertiar) im zentralen Apennin: Eclogae Geologicae Helvetiae, v. 29, p. 1–149. In addition to the microscopic Foraminifera, Italian geologists have made use of the even smaller "nannoplankton," which come from floating photosynthetic algae, in dating the Apennine sediments. For example, see Monechi, S., 1977, Upper Cretaceous and early Tertiary nannoplankton from the *Scaglia Umbra* Formation (Gubbio, Italy): Rivista Italiana di Paleontologia, v. 83, p. 759–802.

5. An example of Bonarelli's later work, incorporating the new dates, is the very beautiful and influential map of the 1:100,000 Gubbio sheet of the Geological Map of Italy (Bonarelli, G., Principi, P., Pilotti, C., Scarsella, F., Lipparini, T., Moretti, A., Selli, R., and Manfredini, M., 1952, Carta Geologica d'Italia, Foglio 116 "Gubbio": Roma, Servizio Geologico d'Italia), which clearly portrays the anticlinal folds from Gubbio to Furlo.

6. This whole episode is told with clarity and sympathy by Roberto Signorini in his introduction to the posthumous publication of Bonarelli's Perugia map (1967, op. cit., p. 15–22).

7. Siever, R., 1988, Sand: New York, W. H. Freeman (Scientific American Library), 237 p.

8. Valloni, R., and Zuffa, G. G., 1984, Provenance changes for arenaceous formations of the Northern Apennines, Italy: Geological Society of America Bulletin, v. 95, p. 1035–1039; Cibin, U., Spadafora, E., Zuffa, G. G., and Castellarin, A., 2001, Continental collision history from arenites of episutural basins in the Northern Apennines, Italy: Geological Society of America Bulletin, v. 113, p. 4–19; Valloni, R., and Basu, A., eds., 2002, Quantitative provenance studies in Italy, Memorie Descrittive della Carta Geologica d'Italia, v. 61, 144 p.

9. I heard this story long ago, and no longer remember from whom. But Arnold Bouma, who was an early student of Kuenen, has told me it is correct. Bouma was himself to play a critical role in the turbidite revolution, explaining the significance of structures within turbidite beds in terms of the flow regime of the turbidity current: Bouma, A. H., 1962, Sedimentology of some flysch deposits: Amsterdam, Elsevier, 168 p. "Bouma sequences" have become a critical tool in understanding turbidites.

10. Kuenen, P. H., and Migliorini, C. I., 1950, Turbidity currents as a cause of graded bedding: Journal of Geology, v. 58, p. 91–127. Bramlette, M. N., and Bradley, W. H., 1940, Geology and biology of North Atlantic deep-sea cores, part I: U.S. Geological Survey Professional Paper, v. 196-A, p. 1–34.

11. Potter, P. E., and Pettijohn, F. J., 1963, Paleocurrents and basin analysis: New York, Academic Press, 296 p. (2nd edition, 1977, Berlin, Springer-Verlag, 425 p.).

12. ten Haaf, E., 1959, Graded beds of the Northern Apennines [PhD thesis], University of Groningen; ten Haaf, E., 1964, Flysch formations in the Northern Apennines, in Bouma, A. H., and Brouwer, A., eds., Turbidites (Developments in Sedimentology, v. 3): Amsterdam, Elsevier, p. 127–136.

13. Mutti, E., and Ricci-Lucchi, F., 1972, Le torbiditi dell'Appennino Settentrionale: Introduzione all'analisi di facies: Società Geologica Italiana Memorie, v. 11, p. 161–199.

14. Ricci-Lucchi, F., 2003, Turbidites and foreland basins; an Apenninic perspective: Marine and Petroleum Geology, v. 20, p. 727–732.

15. Ricci-Lucchi, F., 1995, Sedimentographica: Photographic atlas of sedimentary structures: New York, Columbia University Press, 255 p.

Chapter 11. Paroxysm in the Apennines

1. Procopius, ca. 560, History of the wars, v. 3 (The Gothic War): Cambridge, Mass., Harvard University Press (1919, 1968), Book VI, ch. xi; Alvarez, W., 1999, Drainage on evolving fold-thrust belts: a study of transverse canyons in the Apennines: Basin Research, v. 11, p. 267–284.

2. Merla, G., 1951, Geologia dell'Appennino settentrionale: Società Geologica Italiana Bollettino, v. 70, p. 95–382.

3. Here there is potential for confusion. There is a *Ligurian* Sea today, lying west of northern Italy and separating Corsica from southern France. The Liguride Ocean is different—it no longer exists, and it is known just from the scraps of ocean crust and oceanic sediments that were driven up over Italy when the Liguride Ocean was squeezed shut. The Italian geologists chose the name "Liguride" because this former ocean once lay roughly where the present Ligurian Sea subsequently opened up.

4. Here is another possible source of confusion: *Scaglia* means "scale" or "flake," and the *Scaglia* limestone has that name because you can knock off flakes with a hammer. *Scagliose* means "scaly," and the *argille scagliose* have been torn into innumerable scaly clay fragments by their transport across Italy. *Scaglia* and *Argille scagliose* are completely different and unrelated kinds of rocks.

5. Maxwell, J. C., 1959, Orogeny, gravity tectonics, and turbidites in the Monghidoro area, northern Apennine Mountains, Italy: New York Academy of Sciences Transactions, v. 21, p. 269–280.

6. Roberto Signorini, a central figure in the next chapter, was the discoverer of the huge slab of rock at Monghidoro. Roaming the Apennines in the 1930s with a clear understanding of graded beds (although not yet of turbidity flows), he recognized that this was a "vast zone of overturned strata": Signorini, R., 1938, Una vasta zona a strati rovesciati tra l'Idice e il Setta nell'Appennino Bolognese: Società Geologica Italiana Bollettino, v. 57, p. 139–154.

7. Hsu, K. J., 1967, Origin of large overturned slabs of Apennines, Italy: American Association of Petroleum Geologists Bulletin, v. 51, p. 65–72.

8. An early summary of the Epiligurides was given by Sestini, G., 1970, Development of the Northern Apennines geosyncline—sedimentation of the late geosynclinal stage: Sedimentary Geology, v. 4, p. 445–479. A recent map together with detailed field data has been contributed by Cerrina Feroni, A., Ottria, G., Martinelli, P., Martelli, L., and Catanzariti, R., 2002, Carta geologico-strutturale dell'Appennino emiliano-romagnolo: Bologna, Regione Emilia-Romagna, Servizio Geologico, Sismico, e dei Suoli.

9. A geological guide and map for Canossa and nearby places is available: Regione Emilia-Romagna, Servizio Geologico, Sismico, e dei Suoli, 2004, Geology and environments: Itineraries between Canossa and Quattro Castella (1:15,000).

10. Merla, op. cit.

11. Ibid., p. 286–287.

12. The debate is summarized by Abbate, E., Bortolotti, V., Passerini, P., and Sagri, M., 1970, Introduction to the geology of the Northern Apennines: Sedimentary Geology, v. 4, p. 207–249, esp. p. 235–245.

13. Maxwell, op. cit.; Page, B. M., 1963, Gravity tectonics near Passo della Cisa, Northern Apennines, Italy: Geological Society of America Bulletin, v. 74, p. 655–672; Hsu, K. J., op. cit.

14. The great centers for study of the Northern Apennines were at first the universities at Florence and Pisa, soon followed by Bologna, Camerino, Modena, Parma, Padua, Pavia, Perugia, Siena, and Urbino.

15. Ernesto Abbate, Valerio Bortolotti, Pietro Passerini, Mario Sagri, and Giuliano Sestini singly and in various combinations contributed ten major papers reviewing twenty years of Apennine research triggered by Merla's 1951 synthesis: Sestini, G., ed., 1970, Development of the Northern Apennines geosyncline: Sedimentary Geology, v. 4, no. 3–4, p. 203–647. For years the 1970 volume by the young Florentines was the bible of Apennine geologists, and the copy they gave me in 1971 is badly tattered and full of scrawled notes.

16. Abbate, E., and Sagri, M., 1970, Development of the Northern Apennines geosyncline: the eugeosynclinal sequences: Sedimentary Geology, v. 4, p. 251–340.

17. Vai, G. B., and Martini, I. P., eds., 2001, Anatomy of an orogen: the Apennines and adjacent Mediterranean Basins: Dordrecht, Kluwer Academic Publishers, 632 p.

18. In the Apennines we discovered that compressional shortening of the beds takes place in part by dissolving and removing some of the limestone, leaving clay-coated seams called solution cleavage: Alvarez, W., Engelder, T., and Lowrie, W., 1976, Formation of spaced cleavage and folds in brittle limestone by dissolution: Geology, v. 4, p. 698–701; Alvarez, W., Engelder, T., and Geiser, P. A., 1978, Classification of solution cleavage in pelagic limestones: Geology, v. 6, p. 263–266.

19. Bally, A. W., Gordy, P. L., and Stewart, G. A., 1966, Structure, seismic data, and orogenic evolution of southern Canadian Rocky Mountains: Bulletin of Canadian Petroleum Geology, v. 14, p. 337–381; Boyer, S. E., and Elliott, D., 1982, Thrust systems: American Association of Petroleum Geologists Bulletin, v. 66, p. 1196–1230.

20. Bally et al., op. cit.; Boyer and Elliott, op. cit.

21. Occasionally complications arise, and new ramps do not form in the usual order. These are called out-of-sequence thrusts, and they make the geology more difficult and interesting to figure out.

22. Bally, A. W., Burbi, L., Cooper, C., and Ghelardoni, R., 1986, Balanced sections and seismic reflection profiles across the Central Apennines: Società Geologica Italiana Memorie, v. 35, p. 257–310; Alvarez, W., 1999, Drainage on evolving

fold-thrust belts: a study of transverse canyons in the Apennines: Basin Research, v. 11, p. 267–284.

23. Of course you also need to know the velocity of sound in the rock, which is the first of many complications that make the seismic technique subtle and costly.

24. The most recent collection of deep profiles based on seismic and well data is by Casero, P., 2004, Structural setting of petroleum exploration plays in Italy, in Crescenti, U., D'Offizi, S., Merlini, S., and Sacchi, L., eds., Geology of Italy: Rome, Società Geologica Italiana, p. 189–199.

25. Roberto Colacicchi and Paolo Pialli are both in the photograph on page 105.

26. Colacicchi, R., Passeri, L., and Pialli, G., 1970, Nuovi dati sul Giurese umbro-marchigiano ed ipotesi per un suo inquadrimento regionale: Società Geologica Italiana Memorie, v. 9, p. 839–874.

27. In addition to Giusy, Paolo's group included Giorgio Minelli, Marco Menichetti, Massimiliano Barchi, Francesco Brozzetti, the Dutch geologist Arnoud de Feyter, and the Brazilian geologist João Keller. Other pioneers of Italian structural geology include Gaetano Giglia and Tonino Decandia.

28. The results from CROP–03 are presented in overview by Barchi, M. R., Minelli, G., and Pialli, G., 1998, The CROP–03 profile: A synthesis of results on deep structures of the Northern Apennines: Società Geologica Italiana Memorie, v. 52, p. 383–400, and by Decandia, F. A., Lazzarotto, A., Liotta, D., Cernobori, L., and Nicolich, R., 1998, The CROP–03 traverse: Insights on post-collisional evolution of Northern Apennines: Società Geologica Italiana Memorie, v. 52, p. 427–439, with details in other papers in the same volume. See also Pauselli, C., Barchi, M. R., Federico, C., Magnani, M. B., and Minelli, G., 2006, The crustal structure of the Northern Apennines (central Italy); an insight by the CROP03 seismic line: American Journal of Science, v. 206, p. 428–450; Barchi, M. R., Pauselli, C., Chiarabba, C., di Stefano, R., and Federico, C., 2006, Crustal structure, tectonic evolution and seismogenesis in the Northern Apennines (Italy): Bolletino di Geofisica Teorica e Applicata, v. 47, p. 1–21. The entire CROP seismic program in Italy and adjoining seas is summarized by Scrocca, D., Doglioni, C., Innocenti, F., Manetti, P., Mazzotti, A., Bertelli, L., Burbui, L., and D'Offizi, S., eds., 2003, CROP Atlas: seismic reflection profiles of the Italian crust: Memorie Descrittive della Carta Geologica d'Italia, v. 62, and by Finetti, I. R., ed., 2005, CROP Project: deep seismic exploration of the central Mediterranean and Italy (v. 1, text; v. 2, maps): Amsterdam, Elsevier.

29. Bally et al., op. cit. 1986.

Chapter 12. Tearing the Apennines Apart

1. Wonderful new digital elevation models of most of the world, based on direct radar measurements of elevations, have just become available from the Shuttle Radar Topographic Mission (SRTM), a joint project of NASA and the space agencies of Italy and Germany. SRTM flew on the Space Shuttle Atlantis in 2000, and the data can now be downloaded from the Web.

2. Signorini, R., 1938, Una vasta zona a strati rovesciati tra l'Idice e il Setta nell'Appennino Bolognese: Società Geologica Italiana Bollettino, v. 57,

p. 139–154; Signorini, R., 1945, Sull'inizio della sedimentazione arenacea nell'Appennino Centrale e Settentrionale: Società Geologica Italiana Bollettino, v. 64, p. 27–30.

3. Signorini, R., 1946, Un carattere strutturale frequente nell'Italia centrale: Società Geologica Italiana Bollettino, v. 65, p. 17–21.

4. Prof. Baccetti, quoted by Moretti, A., 1980, Roberto Signorini: Società Geologica Italiana Memorie, v. 21, p. 5–15 (transl.); Merla, G., 1980, Ricordo di Roberto Signorini: Società Geologica Italiana Memorie, v. 21, p. 17–21.

5. Goguel, J., 1954, Rapport sur l'attribution du prix Prestwich à MM. G. Merla et L. Trevisan: Compte Rendu Sommaire des Séances de la Société Géologique de France, v. 9–10, p. 211–216.

6. Trevisan, L., 1951, Sul complesso sedimentario del Miocene superiore e Pliocene della Val di Cecina e sui movimenti tettonici tardivi in rapporto ai giacimenti di lignite e di salgemma: Società Geologica Italiana Bollettino, v. 70, p. 65–78.

7. Elter, P., Giglia, G., Tongiorgi, M., and Trevisan, L., 1975, Tensional and compressional areas in the recent (Tortonian to present) evolution of the Northern Apennines: Bollettino di Geofisica Teorica ed Applicata, v. 17, p. 3–18. I am aware that a number of Italian geologists now doubt the existence of a moving extensional front. That will be decided in the future. I tell this part of the story on the basis of the coupled, moving fronts, because this idea has been so influential for so many years.

8. Mazzanti, R., and Trevisan, L., 1978, Evoluzione della rete idrografica nell'Appennino centro-settentrionale: Geografia Fisica e Dinamica Quaternaria, v. 1, p. 55–62.

9. Long before the concepts were there to explain the pattern of the Tyrrhenian rivers, Giovanni Merla made a detailed study of the Tiber: 1938, Il Tevere. Monografia idrologica, geologia e permeabilità dei terreni del bacino, Servizio Idrografico (Roma), Pubblicazioni, v. 22, 129 p.

10. Purgatory, Canto XIV, verses 22–48. Thanks to Antonio Lazzarotto for telling me about this vivid geological metaphor from Dante.

11. If you look at a map of the region of Marche that shows the rivers and does not show the mountains, it looks as if the rivers are wandering across a gentle coastal plain. You simply cannot tell from the drainage map that the mountains are there. This is why Milly and I, driving through this region for the first time in the zero visibility of a winter storm, with only an old Esso road map to guide us, were so surprised when the clouds cleared for a moment and we saw that we were in the mountainous gorge that we later learned is called Furlo.

12. Geologists have long had two explanations for rivers that cut through mountain barriers. One explanation is that the river is "antecedent"—older than the mountain—so that the river just kept cutting down through the mountain as the mountain was pushed up. This does not work for the Apennines, because they are made of marine sedimentary rocks, so the anticlinal folds must have emerged right out of the sea, and there could have been no river there beforehand. The other explanation is that the river is "superposed," meaning that the mountains were there first but were completely buried by younger sediment. The river would have wandered across the flat top of the younger sediments and then have been

let down onto the buried mountain as the whole region was uplifted and the river eroded down into the young sediments. This also does not work for the Adriatic rivers, because the folds are so young—the youngest are forming even now—that there would be no time to bury them under still younger sediments, and indeed no such younger sediments have been found.

13. Alvarez, W., 1999, Drainage on evolving fold-thrust belts: a study of transverse canyons in the Apennines: Basin Research, v. 11, p. 267–284. Among the interesting things I found was that in many cases the rivers do cut through the anticlines, but in some cases a river detours around an anticline. That would happen if the fold emerged from the sea, as an island, before the new downstream tract of the river could get across it. This seems to me a worthy corollary of the concept of Mazzanti and Trevisan—that you can look at a map of the mountains, see whether the rivers transect them or avoid them, and from that pattern you can decide whether there were islands off the Adriatic shore at some time in the past.

14. Fazzini, P., and Mantovani, M. P., 1965, La geologia del gruppo di M. Subasio: Società Geologica Italiana Bollettino, v. 84, no. 3, p. 71–142.

15. Barchi, M. R., de Feyter, A., Magnani, M. B., Minelli, G., Pialli, G., and Sotera, M., 1998, Extensional tectonics in the Northern Apennines (Italy): Evidence from the CROP 03 deep seismic reflection line: Società Geologica Italiana Memorie, v. 52, p. 527–538; Barchi, M. R., Minelli, G., and Pialli, G., 1998, The CROP-03 profile: A synthesis of results on deep structures of the Northern Apennines: Società Geologica Italiana Memorie, v. 52, p. 383–400; Barchi, M. R., Paolacci, S., Pauselli, C., Pialli, G., and Merlini, S., 1999, Geometria delle deformazioni estensionali recenti nel bacino del'Alta Val Tiberina fra S. Giustino Umbro e Perugia; evidenze geofisiche e considerazioni geologiche: Società Geologica Italiana Bollettino, v. 118, p. 617–625.

16. Boncio, P., Brozzetti, F., and Lavecchia, G., 2000, Architecture and seismotectonics of a regional low-angle normal fault zone in central Italy: Tectonics, v. 19, p. 1038–1055.

17. Lavecchia, G., Brozzetti, F., Barchi, M., Menichetti, M., and Keller, J. V. A., 1994, Seismotectonic zoning of east-central Italy deduced from an analysis of the Neogene to present deformations and related stress fields: Geological Society of America Bulletin, v. 106, p. 1107–1120; Boncio, P., and Lavecchia, G., 1999, I terremoti di Colfiorito (Appennino umbro-marchigiano) del settembre-ottobre 1997; contesto tettonico e prime considerazioni sismogenetiche: Società Geologica Italiana Bollettino, v. 118, p. 229–236.

Chapter 13. Salt Crisis

1. de Feyter, A. J., 1991, Gravity tectonics and sedimentation of the Montefeltro, Italy: Geologica Ultraiectina, v. 35, p. 64–69.

2. Martini, I. P., and Sagri, M., 1993, Tectono-sedimentary characteristics of Late Miocene-Quaternary extensional basins of the Northern Apennines, Italy: Earth-Science Reviews, v. 34, p. 197–233.

3. Evaporites form the third great group of sedimentary rocks, along with "clastic"

(broken) rocks like sandstones and conglomerates, and "carbonates," like limestones, that form mostly from the shells of living plants and animals.

4. Moorehead, A., 1960, The White Nile: New York, Harper and Brothers, p. 1.

5. Said, R., 1981, The geological evolution of the River Nile: New York, Springer-Verlag, 151 p.

6. Egypt is part of the very ancient core of the African continent, so old and stable that geologists would not expect the continent itself to rise or subside very much. So the deep, buried Nile canyon could not be explained by invoking uplift, then erosion, and then subsidence.

7. Barber, P. M., 1981, Messinian subaerial erosion of the proto-Nile Delta: Marine Geology, v. 44, p. 253–272.

8. Chumakov, I. S., 1973, Pliocene and Pleistocene deposits of the Nile Valley in Nubia and Upper Egypt, Initial Reports of the Deep Sea Drilling Project, v. 13, p. 1242–1243; Barber, op. cit.

9. Other deeply incised river channels dating from the Late Miocene were found in Libya and southern France: Barr, F. T., and Walker, B. R., 1973, Late Tertiary channel system in northern Libya and its implications on Mediterranean sea level changes, Initial Reports of the Deep Sea Drilling Project, v. 13, part 2, p. 1244–1251; Clauson, G., 1973, The eustatic hypothesis and the pre-Pliocene cutting of the Rhône Valley, Initial Reports of the Deep Sea Drilling Project, v. 13, part 2, p. 1251–1256.

10. The Deep-Sea Drilling Project drilled 624 sites in oceans around the world from 1968 to 1983. Its successor, the Ocean Drilling Project, with a more advanced ship, drilled 650 more sites between 1985 and 2003. The Integrated Ocean Drilling Project began in 2004 and continues to harvest the archives of Earth history stored on the bottom of the ocean. The thick volumes of technical reports fill shelf after shelf in geology libraries.

11. The scientists on Leg 13 were Maria Bianca Cita (University of Milan, Italy), Paulian Dumitrica (Geological Institute of Bucharest, Romania), Jennifer M. Lort (Cambridge University, England), Wolf Maync (Bern, Switzerland), Wladimir D. Nesteroff (University of Paris VI), Guy Pautot (Centre Océonologique de Bretagne, Brest, France), Herbert Stradner (Geological Survey of Austria, Vienna), and Forese Carlo Wezel (University of Catania, Italy).

12. Ryan, W. B. F., Hsü, K. J., et al., 1973, Initial Reports of the Deep Sea Drilling Project: Washington, D.C., U.S. Government Printing Office, v. 13 (part 1, 514 p.; part 2, 1447 p.); Hsü, K. J., 1983, The Mediterranean was a desert: Princeton, N. J., Princeton University Press, 197 p. For Bill Ryan's account, see Ryan, W., and Pitman, W., 1998, Noah's Flood: The new scientific discoveries about the event that changed history: New York, Simon and Schuster, 319 p., ch. 7.

13. Hsü, K. J., Ryan, W. B. F., and Schrieber, B. C., 1973, Petrography of a halite sample from Hole 134—Balearic Abyssal Plain, in Ryan, W. B. F., and Hsü, K. J., eds., Initial Reports of the Deep Sea Drilling Project, v. 13: Washington, D.C., U.S. Government Printing Office, p. 708–711.

14. Gypsum has the chemical formula $CaSO_4 \cdot 2H_2O$; anhydrite is $CaSO_4$.

15. This was not just a strange coincidence. Geologists much earlier had *defined* the Miocene-Pliocene boundary as the moment when sudden changes took

place in the Mediterranean, although they did not know what had caused those changes.

16. The evaporation at the surface of the Mediterranean and the replacement water flowing in through the Strait of Gibraltar produce a complicated circulation. The surface waters of the Mediterranean, made dense by evaporation, sink to the bottom and are flushed out through the strait, where they flow westward *beneath* the east-flowing Atlantic water entering the Mediterranean.

17. As water from the Gibraltar waterfall surged across the desiccated Mediterranean Sea floor, it may have catastrophically eroded the deep cleft through the M-reflector at Site 126 that the *Challenger* scientists used to find out what was below the evaporites. The water would have cut this channel on its way down to the very deep trench south of Crete (Hsü, K. J., Cita, M. B., and Ryan, W. B. F., 1973, The origin of the Mediterranean evaporites, in Ryan, W. B. F., and Hsü, K. J., eds., Initial Reports of the Deep Sea Drilling Project, v. 13: Washington, D.C., U.S. Government Printing Office, p. 1203–1231, esp. p. 1218).

18. Hsü, K. J., Ryan, W. B. F., and Cita, M. B., 1973, Late Miocene desiccation of the Mediterranean: Nature, v. 242, p. 240–244; Hsü, K. J., Cita, M. B., and Ryan, W. B. F., 1973, op. cit.

19. And soon they also learned about the filled canyons of Messinian age in Libya, southern France, and elsewhere around the Mediterranean (see note 9 for this chapter).

20. For arguments against the full sea-level drawdown, see Roveri, M., Bassetti, M. A., and Ricci-Lucchi, F., 2001, The Mediterranean Messinian salinity crisis: an Apennine foredeep perspective: Sedimentary Geology, v. 140, p. 201–214.

21. Clauzon, G., Suc, J.-P., Gautier, F., Berger, A., and Loutre, M.-F., 1996, Alternate interpretation of the Messinian salinity crisis; controversy resolved?: Geology, v. 24, p. 363–366; detailed dates are given by Krijgsman, W., Hilgen, F. J., Raffi, I., Sierro, F. J., and Wilson, D. S., 1999, Chronology, causes and progression of the Messinian salinity crisis: Nature, v. 400, p. 652–655.

22. There is far more gypsum and salt down there than you could get from one "Mediterranean-full" of seawater, so Atlantic seawater must have continually leaked into the Mediterranean desert during the Messinian salinity crisis— enough to yield gypsum and salt through evaporation, but not enough to keep the Mediterranean full of water.

23. As noted in ch. 5, Steno's recognition of the Pliocene marine transgression was pointed out by Rodolico, F., 1971, Niels Stensen, founder of the geology of Tuscany, in Scherz, G., ed., Dissertations on Steno as geologist: Odense, Denmark, Odense Universitetsforlag, p. 237–243.

24. Ryan, W., and Pitman, W., op. cit. When Milly and I were at Lamont in the 1970s, she worked for Walter Pitman and spent a lot of time in the library tracking down versions of the flood legends that eventually went into the Ryan-Pitman book. Recent exploration of the floor of the Black Sea has provided remarkable support for the Ryan-Pitman hypothesis (Ballard, R. D., Coleman, D. F., and Rosenberg, G. D., 2000, Further evidence of abrupt Holocene drowning of the Black Sea shelf: Marine Geology, v. 170, p. 253–261). On the other hand studies of the sea floor just outside the Dardanelles Strait,

which connects the Mediterranean and the Black Sea, seem to contradict the hypothesis (Aksu, A. E., Mudie, P. J., Rochon, A., Kaminski, M. A., Abrajano, T., and Yasar, D., 2002, Persistent Holocene outflow from the Black Sea to the Eastern Mediterranean contradicts Noah's Flood hypothesis: GSA Today, v. 12, p. 4–10). This is still very much a cutting-edge research topic. For a broader view of repeated Black Sea flooding during glacial times, see Ryan, W. B. F., Major, C. O., Lericolais, G., and Goldstein, S. L., 2003, Catastrophic flooding of the Black Sea: Annual Review of Earth and Planetary Sciences, v. 31, p. 525–554, and for a series of papers debating the Black Sea flood hypothesis, see Yanko-Hombach, V., Gilbert, A. S., Panin, N., and Dolukhanov, P. M., 2007, The Black Sea flood question: changes in coastline, climate and human settlement: Dordrecht, Springer, p. 971.

Chapter 14. Beyond Plate Tectonics

1. Glen, W., 1982, The road to Jaramillo: Stanford, Stanford University Press, 459 p.; Oreskes, N., ed., 2001, Plate tectonics: an insider's history of the modern theory of the Earth: Boulder, Colo., Westview Press, 424 p.
2. Nairn, A. E. M., and Westphal, M., 1968, Possible implications of the palaeomagnetic study of Late Paleozoic igneous rocks of northwestern Corsica: Palaeogeography, Palaeoclimatology, Palaeoecology, v. 5, p. 179–204; De Jong, K. A., Manzoni, M., and Zijderveld, J. D. A., 1969, Palaeomagnetism of the Alghero trachyandesites: Nature, v. 224, p. 67–69; Zijderveld, J. D. A., De Jong, K. A., and van der Voo, R., 1970, Rotation of Sardinia: Palaeomagnetic evidence from Permian rocks: Nature, v. 226, p. 933–934.
3. Alvarez, W., 1972, Rotation of the Corsica-Sardinia microplate: Nature Physical Science, v. 235, p. 103–105.
4. Calabria is one of the few places in the world where you can walk around on rocks that once lay near the base of the continental crust, before uplift and erosion brought them up to the surface: Schenk, V., 1988, The exposed crustal cross section of southern Calabria, Italy: structure and evolution of a segment of Hercynian crust, in Salisbury, M. H., and Fountain, D. M., eds., Exposed cross sections of the continental crust: Dordrecht, Kluwer, p. 21–42; Piluso, E., and Morten, L., 2004, Hercynian high temperature granulites and migmatites from the Catena Costiera, northern Calabria, southern Italy: Periodico di Mineralogia, v. 73, p. 159–172.
5. Cortese, E., 1895, Descrizione geologica della Calabria: Memorie Descrittive della Carta Geologica d'Italia, v. 9, p. 1–310.
6. Limanowski, M., 1913, Wielka plaszczowina kalabryjska—Die grosse Kalabrische Decke: Bulletin International de l'Académie des Sciences de Cracovie, Classe des Sciences Mathématiques et Naturelles, Sér. A: Sciences Mathématiques, v. 6a, p. 370–385; Quitzow, H. W., 1935, Der Deckenbau des Kalabrischen Massivs und seiner Randgebiete: Abhandlungen der Akademie der Wissenschaften in Goettingen, Mathematisch-Physikalische Klasse, v. 13, p. 63–179.
7. The history of geological exploration in Calabria is reviewed in great detail by Ogniben, L., 1973, Schema geologico della Calabria in base ai dati odierni:

Geologica Romana, v. 12, p. 243–585. A very influential map and synthesis were published by Bonardi, G., Colonna, V., Dietrich, D., Giunta, G., Liguori, V., Lorenzoni, S., Paglionico, A., Perrone, V., Piccarreta, G., Russo, M., Scandone, P., Zanettin Lorenzoni, A., and Zuppetta, A., 1976, L'Arco calabro-peloritano, Carta Geologica 1:500.000, Società Geologica Italiana, 68° Congresso, and by Amodio-Morelli, L., Bonardi, G., Colonna, V., Dietrich, D., Giunta, G., Ippolito, F., Liguori, V., Lorenzoni, S., Paglionico, A., Perrone, V., Piccarreta, G., Russo, M., Scandone, P., Zanettin-Lorenzoni, E., and Zuppetta, A., 1976, L'arco Calabro-Peloritano nell'orogene Appenninico-Maghrebide: Società Geologica Italiana Memorie, v. 17, p. 1–60. For a recent synthesis, see Bonardi, G., Cavazza, W., Perrone, V., and Rossi, S., 2001, Calabria-Peloritani Terrane and Northern Ionian Sea, in Vai, G. B., and Martini, I. P., eds., Anatomy of an orogen: the Apennines and adjacent Mediterranean Basins: Dordrecht, Kluwer Academic Publishers, p. 287–306.

8. Alvarez, W., Cocozza, T., and Wezel, F. C., 1974, Fragmentation of the Alpine orogenic belt by microplate dispersal: Nature, v. 248, p. 309–314. We were not the first to reach this conclusion, for a French group had published a similar reconstruction a couple of years earlier (Haccard, D., Lorenz, C., and Grandjacquet, C., 1972, Essai sur l'évolution tectogénétique de la liason Alpes-Apennins (de la Ligurie a la Calabrie): Società Geologica Italiana Memorie, v. 11, p. 309–341).

9. Ryan, W. B. F., Hsü, K. J., et al., 1973, Initial Reports of the Deep Sea Drilling Project: Washington, D.C., U.S. Government Printing Office, v. 13, part 1, p. 444–445. We knew about this in detail because Forese was one of the Leg 13 shipboard scientists.

10. Alvarez, Cocozza, and Wezel, op. cit.

11. The new drilling ship had replaced Glomar Challenger when the Deep-Sea Drilling Project was succeeded by the Ocean Drilling Program. Kim Kastens and Jean Mascle were the Chief Scientists for Leg 107 in the Tyrrhenian Sea.

12. Kastens, K., Mascle, J., et al., 1988, ODP Leg 107 in the Tyrrhenian Sea: Insights into passive margin and back-arc basin evolution: Geological Society of America Bulletin, v. 100, p. 1140–1156.

13. Elter, P., Giglia, G., Tongiorgi, M., and Trevisan, L., 1975, Tensional and compressional areas in the recent (Tortonian to present) evolution of the Northern Apennines: Bollettino di Geofisica Teorica ed Applicata, v. 17, p. 3–18. This paper is discussed in ch. 12.

14. Doglioni, C., Merlini, S., and Cantarella, G., 1999, Foredeep geometries at the front of the Apennines in the Ionian Sea (central Mediterranean): Earth and Planetary Science Letters, v. 168, p. 243–254.

15. Elter et al., op. cit.

16. Reutter, K. J., Giese, P., and Closs, H., 1980, Lithospheric split in the descending plate: observations from the Northern Apennines: Tectonophysics, v. 64, p. T1–T9; Castellarin, A., Colacicchi, R., Praturlon, A., and Cantelli, C., 1982, The Jurassic-Lower Pliocene history of the Anzio-Ancona Line (Central Italy): Società Geologica Italiana Memorie, v. 24, p. 325–336; Malinverno, A., and Ryan, W. B. F., 1986, Extension in the Tyrrhenian Sea and shortening in the

Apennines as a result of arc migration driven by sinking of the lithosphere: Tectonics, v. 5, p. 227–245. These papers all agree on the general cause of Apennine deformation; they differ on the geometrical details.

17. The earthquake locations I have shown on this map come from the catalog of the National Earthquake Information Center of the U.S. Geological Survey, from 1973 through 2006 (magnitude 4 and greater; depth greater than one hundred kilometers). In the next few years much better information should become available from the CAT/SCAN (Calabria-Apennine-Tyrrhenian/Subduction-Collision-Accretion Network) project of Lamont-Doherty Earth Observatory and the Italian Istituto Nazionale di Geofisica e Vulcanologia, in which many seismographs on land and on the ocean floor will monitor Tyrrhenian earthquakes up close.

18. Kay, R. W., and Mahlburg-Kay, S., 1991, Creation and destruction of lower continental crust: International Journal of Earth Sciences, v. 80, p. 259–278. Paolo Pialli, Giorgio Minelli, and I pointed out that this kind of delamination and rollback of the lower continental crust, resulting from conversion of basaltic rock to eclogite, could explain the coupled fronts of compression and extension in the Apennines: Pialli, G., Alvarez, W., and Minelli, G., 1995, Geodinamica dell'Appennino settentrionale e sue ripercussioni nella evoluzione tettonica miocenica: Studi Geologici Camerti, volume speciale no. 1, p. 523–536; Pialli, G., and Alvarez, W., 1997, Tectonic setting of the Miocene Northern Apennines: the problem of contemporaneous compression and extension, in Montanari, A., Odin, G. S., and Coccioni, R., eds., Miocene stratigraphy: an integrated approach: Amsterdam, Elsevier, p. 167–185.

19. Earthquakes occur in the sinking slab under the southeastern Tyrrhenian Sea, presumably because this slab is made of old oceanic crust of the Ionian Sea. Sinking ocean crust, cool and become brittle, can fracture and generate earthquakes, which do let us see into the deep Earth. This is not the case under the Apennines. If there is delaminating lower continental crust under the Apennines, it was never close enough to the surface to cool and become brittle. Still warm and ductile, it would deform by flowing and would never fracture and generate earthquakes, so we cannot see it in the way we can see the Tyrrhenian slab.

20. Here's how seismic tomography works. Every year there are many earthquakes all over the world, and each is recorded at many seismographs, also spread over the entire Earth. This makes for a huge number of travel paths followed by seismic waves, linking each earthquake to each seismograph, and one path or another travels through most parts of the Earth's interior. If there is a cold, dense body down there somewhere, like a delaminated chunk of lower continental crust, it will let the seismic waves travel a little faster for that part of their journey, and they will arrive at a seismograph just a bit earlier than waves that did not go through any cold, dense, high-velocity rock. So if you take all the arrival times, and note how many seconds they came in ahead of or behind the expected arrival time, and then run them all through a sophisticated computer program, you can find out where there are dense patches at depth. In practice, of course, it is subtle, complex, and mathematically very difficult, but as a result

of this tomographic technique we now have wonderful images of the Earth's interior.

21. Spakman, W., 1988, Upper mantle delay time tomography with an application to the collision zone of the Eurasian, African and Arabian plates: Geologica Ultraiectina, v. 53, p. 1–200.

22. Peccerillo, A., 1985, Roman comagmatic province (central Italy); evidence for subduction-related magma genesis: Geology, v. 13, p. 103–106; Serri, G., Innocenti, F., and Manetti, P., 1993, Geochemical and petrological evidence of the subduction of delaminated Adriatic continental lithosphere in the genesis of the Neogene-Quaternary magmatism of central Italy: Tectonophysics, v. 223, p. 117–147; Peccerillo, A., 2005, Pilo-Quaternary volcanism in Italy: petrology, geochemistry, geodynamics: Berlin, Springer, 365 p.

23. This is the RETREAT Project: earth.geology.yale.edu/RETREAT.

Epilogue

1. Notable publications on the occasion of the Thirty-second International Geological Congress include Crescenti, U., D'Offizi, S., Merlini, S., and Sacchi, L., eds., 2004, Geology of Italy: Special volume of the Italian Geological Society for the IGC 32 Florence–2004: Roma, Società Geologica Italiana, 232 p., and Vai, G. B., and Cavazza, W., eds., 2003, Four centuries of the word Geology: Ulisse Aldrovandi 1603 in Bologna: Bologna, Minerva Ed., 352 p. Information about the congress is available online at www.32igc.org/GeneralProcedings/home.htm.

ILLUSTRATION CREDITS AND NOTES

Page

5 Photograph by Elio Ciol, 1991, Assisi: Milan, Federico Motta. Used by permission.

33 The topography for this map was prepared with the preliminary unedited data files from the Shuttle Radar Topographic Mission. The SRTM is a joint project of the United States, Germany, and Italy, and provides digital elevation models with horizontal resolution of 30 m. The SRTM data were displayed using Jerry Farm's MacDEM software.

40 Old photograph well known in Mazzano, source unknown.

44 This map is based on Fig. 14 and Plate I of Alvarez, W., 1972, The Treia Valley north of Rome: volcanic stratigraphy, topographic evolution, and geological influences on human settlement: Geologica Romana, v. 11, p. 153–176.

75 From a portrait in the Pitti Palace, Florence.

79 In this diagram I have shown four of Steno's six cross sections, and I have redrawn them as block diagrams for clarity. Steps 5 and 6 show a new cavern that collapses in turn. Steno's drawings are reproduced in Steno, N., 1669, De solido intra solidum naturaliter contento dissertationis prodromus: Florence (Transl. J. G. Winter, 1916, The Prodromus of Nicholas Steno's Dissertation, University of Michigan Studies, Humanistic Series, v. 11, pt. 2, New York, Macmillan). However Steven Jay Gould, 1987, Time's Arrow, Time's Cycle: Cambridge, Mass., Harvard University Press, p. 51–59, points out that Steno's original drawing, with two columns of three diagrams each, clearly showing his cyclical view of Earth history, was rearranged into a single column of six diagrams in Winter's 1916 translation.

86 Image courtesy of Ferruccio Farsi, Accademia dei Fisiocritici, Siena.

88 Image courtesy of Ferruccio Farsi.

94 This picture is from a large collection of historical photos of the Festa dei Ceri on the Web: www.ceri.it/ceri/foto/index.htm.

97 These forams are from near the top of the Cretaceous, and they are the species

we called *Globotruncana contusa* when we were doing the paleomagnetic work at Gubbio. Since then micropaleontologists have learned more about the evolution of foraminifera and have modified the genus names, so that this foram is now called *Contusotruncana contusa*. *C. contusa* is unusually large for a *Scaglia* foram, with each individual about 1 mm in diameter, and it is one of the species that perished in the Cretaceous-Tertiary mass extinction.

100 Images Courtesy of Professor Howard J. Spero, University of California, Davis.

140 In this figure I have omitted the complications due to the fact that during the Jurassic, some parts of the *Massiccio* subsided earlier than others (Colacicchi, R., Passeri, L., and Pialli, G., 1970, Nuovi dati sul Giurese umbro-marchigiano ed ipotesi per un suo inquadrimento regionale: Società Geologica Italiana Memorie, v. 9, p. 839–874; Centamore, E., Chiocchini, M., Deiana, G., Micarelli, A., and Pieruccini, U., 1971, Contributo alla conoscenza del Giurassico dell'Appennino Umbro-Marchigiano: Studi Geologici Camerti (Camerino), v. 1, p. 7–89; Cecca, F., Cresta, S., Pallini, G., and Santantonio, M., 1990, Il Guirassico de Monte Nerone (Appenninio marchigiano, Italia Centrale): biostratigrafia, litostratigrafia ed evoluzione paleogeografica, in Pallini, G., Cecca, F., Cresta, S., and Santantonio, M., eds., Fossili, evoluzione, ambiente (Pergola 25–30 ottobre 1987): Pergola, Comitato Centernario Raffaele Piccinini, p. 63–139; Alvarez, W., 1989a, Evolution of the Monte Nerone seamount in the Umbria-Marche Apennines: 1. Jurassic-Tertiary stratigraphy: Società Geologica Italiana Bollettino, v. 108, p. 3–21; 1989b, Evolution of the Monte Nerone seamount in the Umbria-Marche Apennines: 2. Tectonic control of the seamount-basin transition: Società Geologica Italiana Bollettino, v. 108, p. 23–39).

143 For a remarkable piece of detective work, figuring out the life habits of these now-extinct organisms, see Kotake, N., 1989, Paleoecology of the Zoophycos producers: Lethaia, v. 22, p. 327–341.

144 This figure is redrawn from Alvarez, W., Colacicchi, R., and Montanari, A., 1985, Synsedimentary slides and bedding formation in Apennine pelagic limestones: Journal of Sedimentary Petrology, v. 55, p. 720–734, Fig. 6. The stretched and shortened parts of slumps are easily seen in outcrops. Missing beds are harder to detect, but in some cases they may be recognized where fossil zones are missing. At Furlo we were able to identify the undeformed but rotated parts of several slumps because their paleomagnetic declinations were rotated away from north (Alvarez, W., and Lowrie, W., 1984, Magnetic stratigraphy applied to synsedimentary slumps, turbidites, and basin analysis: The Scaglia limestone at Furlo [Italy]: Geological Society of America Bulletin, v. 95, p. 324–336).

145 Simplified after Channell, J. E. T., and Horváth, F., 1976, The African/Adriatic Promontory as a palaeogeographic premise for Alpine orogeny and plate movements in the Carpatho-Balkan Region: Tectonophysics, v. 35, p. 71–101, Fig. 6. There is actually some oceanic crust south of Greece and north of Libya and Egypt, separating the Adriatic continental crust from that of Africa, but the Adriatic crust has moved coherently with Africa.

148 Base map from Wikimedia Commons (http://en.wikipedia.org/wiki/Image: Alpenrelief_02.jpg). Annotations added.

156 *Top*: This photo was taken in 1973 by Professor Stefan Schmid of the University of Basel in Switzerland and is used with his kind permission. The locality is the remote Segnaspass, where in 1848 Arnold Escher first convinced the influential English geologist Roderick Impey Murchison that older rocks could indeed be emplaced over younger ones (Bailey, E. B., 1935 [reprinted 1968], Tectonic essays, mainly Alpine: Oxford, Oxford University Press, p. 46–49; compare Schmid's photo to Bailey's fig. 13, p. 49). Another valuable account of the Glarus discovery and controversy is by Trümpy, R., 1991, The Glarus nappes: a controversy of a century ago, in Müller, D. W., McKenzie, J. A., and Weissert, H., eds., Controversies in modern geology: London, Academic Press, p. 385–404. *Bottom*: This is a famous outcrop, just east of the village of Schwanden. A plaque (translated from the German) reads: "Here, on August 1, 1840, Arnold Escher von der Linth for the first time discovered older rocks (Permian Verrucano) thrust over younger ones (Eocene schist), and thus found a key to Alpine geology."

165 Channell and Horváth, 1976, op. cit. I have simplified these figures by omitting the microcontinents within the Tethyan Ocean. Although very influential when first published, these maps need modification based on newer research; see for example fig. 4C in Bernoulli, D., Manatschal, G., Desmurs, L., and Müntener, O., 2003, Where did Gustav Steinmann see the trinity? Back to the roots of an Alpine ophiolite concept, in Dilek, Y., and Newcomb, S., eds., Ophiolite concept and the evolution of geological thought: Geological Society of America Special Paper no. 373): Boulder, Colo., Geological Society of America, p. 93–110.

169 The *Bisciaro* limestone is a set of clay-poor beds within the generally marly interval between the *Scaglia* limestone and the *Marnoso-arenacea* sandstone. It may be due to a transient bulge on the sea floor beyond the trough due to the weight of the advancing thrust front, explained in ch. 10 and 11 (Montanari, A., and Koeberl, C., 2000, Impact stratigraphy: the Italian record: Lecture notes in Earth science, v. 93: Berlin, Springer, p. 166).

176 Photo used with the permission of the Head of the Dipartimento di Scienze della Terra, University of Florence.

178 Kuenen, P. H., and Migliorini, C. I., 1950, Turbidity currents as a cause of graded bedding: Journal of Geology, v. 58, p. 91–127, plate 3B. The view is 3.5 cm (1.5 in) high.

180 ten Haaf, E., 1959, Graded beds of the Northern Apennines (PhD thesis), University of Groningen, 102 p., Fig. 12.

186 Photo used with the permission of the Head of the Dipartimento di Scienze della Terra, University of Florence.

187 This map is simplified from the opening sketch map (p. 217) of Abbate, E., Bortolotti, V., Passerini, P., and Sagri, M., 1970, Introduction to the geology of the Northern Apennines: Sedimentary Geology, v. 4, p. 207–249, which is the introduction to the first detailed synthesis of Northern Apennine geology: Sestini, G., ed., 1970, Development of the Northern Apennines geosyncline: Sedimentary Geology, v. 4, no. 3–4, p. 203–647. Modern maps show more detail, but the 1970 map still gives a good general picture.

196 This profile is based on the NE half of the lower panel of Barchi, M. R., De Feyter, A., Magnani, M. B., Minelli, G., Pialli, G., and Sotera, M., 1998, Geological interpretation of the CROP 03 NVR seismic profile (stack), Plate 3: Società Geologica Italiana Memorie, v. 52.

204 Società Geologica Italiana Memorie, 1980, v. 21, p. 7.

206 Società Geologica Italiana Memorie, 1993, v. 49, p. 5.

207 Used with the permission of Prof. Michele Marroni, University of Pisa.

208 From Elter, P., Giglia, G., Tongiorgi, M., and Trevisan, L., 1975, Tensional and compressional areas in the recent (Tortonian to present) evolution of the Northern Apennines: Bollettino di Geofisica Teorica ed Applicata, v. 17, p. 3–18.

211 The figure is based on Mazzanti, R., and Trevisan, L., 1978, Evoluzione della rete idrografica nell'Appennino centro-settentrionale: Geografia Fisica e Dinamica Quaternaria, v. 1, p. 55–62, fig. 2. I have added some annotations not in the original. Today we would show the internal thrust structure of each fold, but that was not widely understood when Professor Trevisan drew this sketch.

214 This cross section follows part of section I of 1:50,000 geological map sheet 290 "Cagli," extended southwest to the Tiber Valley. The deep structure is modified from Barchi, M. R., Minelli, G., and Pialli, G., The CROP-03 Profile, and Boncio, P., Brozzetti, F., and Lavecchia, G., 2000, Architecture and seismotectonics of a regional low-angle normal fault zone in central Italy: Tectonics, v. 19, p. 1038–1055.

219 This figure is simplified from Chumakov, I. S., 1973, Pliocene and Pleistocene deposits of the Nile Valley in Nubia and Upper Egypt, Initial Reports of the Deep Sea Drilling Project, v. 13, part 2, p. 1242–1243. I have stretched Chumakov's published profile to make the horizontal and vertical scales equal.

220 This is a MODIS image, owned by NASA and provided by NASA's Visible Earth team (visibleearth.nasa.gov); it was the Earth Picture of the Day for September 29, 2000 (epod.usra.edu). Original in color, not annotated. The nearly black band extending from just north of Aswan to Cairo is the fertile, vegetated floodplain of the Nile. North of Cairo is the inverted triangle of the Nile Delta. The irregularly margined black band south of Aswan is Lake Nasser, the reservoir impounded behind the Aswan Dam.

GLOSSARY

Aa Lava with a blocky surface, formed when the upper part of a lava flow solidifies and cracks apart while the underlying part stays fluid and keeps moving. A Hawaiian word pronounced "AH-ah." (Chap. 3)

Angular unconformity A break in the stratigraphic record, between folded and eroded rocks below the unconformity, and younger rocks, not folded, above it. (Chap. 8)

Anhydrite A white mineral with the composition $CaSO_4$, which precipitates when seawater evaporates, thus becoming very saline. Anhydrite is familiar as plaster of Paris. (Chap. 13)

Anticline A fold where beds close upward, making a structural arch. (Chap. 7)

Ash-fall tuff A volcanic-sedimentary rock made of ash and coarser fragments that were erupted explosively and then fell down out of the atmosphere. (Chap. 3)

Ash-flow tuff A volcanic-sedimentary rock whose ash and volcanic fragments flowed very rapidly downhill, mixed with volcanic gases, often very hot. Also called an ignimbrite. (Chap. 3)

Basalt A black volcanic rock high in magnesium and iron, low in silicon and aluminum, which forms much of the oceanic crust. (Chap. 3)

Basement The igneous or metamorphic rocks underlying the shallow sedimentary rocks. Basement is often the top of the continental or oceanic crust. (Chap. 8)

Bed A layer of sedimentary rock of a particular composition, bounded by layers of other composition. An example typical of the *Scaglia rossa* would be a thick limestone bed bounded above and below by thin clay beds. (Chap. 2)

Bedrock Undisturbed, unaltered rock exposed at the Earth's surface, not covered by soil or by very young sediment. (Chap. 1)

Calcite A white mineral with the composition $CaCO_3$, precipitated by marine organisms to build their shells. Calcite is the mineral of which limestone is made. Calcite can be thought of as $CaO + CO_2$, which makes it clear that limestone stores CO_2 gas in solid form. (Chap. 6)

Chert A very hard, fine-grained sedimentary rock with the same composition as quartz (SiO_2). Chert is usually formed by the dissolution and reprecipitation of the shells of microscopic, single-celled radiolarians, and the spicules, or spines, of sponges, both of which are composed of SiO_2. Chert may occur in beds, or in isolated lumpy nodules within beds of limestone, and in this form it is common in the *Majolica* and parts of the *Scaglia*. (Chap. 6)

Clay A very fine-grained, soft mineral formed by the weathering of bedrock. It is a key ingredient of soil, and when carried to the sea by a river, the fine clay particles can drift for long distances, eventually settling slowly to the sea floor. (Chap. 6)

Compression Horizontal squeezing of part of the crust or sediments, as if they were caught in a vise. (Chap. 11)

Continental crust A body of low-density rocks, mostly igneous and metamorphic, about 35 km thick, resting on top of the mantle, in areas of continents and the shallow seas surrounding them. (Chap. 8)

Core The deepest part of the Earth, from about 2,900 km below the surface down to the center of the Earth at 6,378 km depth. The core is made of metallic iron containing some nickel. The outer shell of the core is molten iron whose convective flow produces the Earth's magnetic field; the inner part is solid iron. (Chap. 2)

Crust The uppermost of the three great entities of which the Earth is made—core, mantle, and crust—extending downward to anywhere from 5 to about 40 km, depending on whether it is oceanic or continental crust. (Chap. 1)

Dip The angle of tilt of a plane (usually a bed or a fault), measured down from the horizontal. A 0° dip means a horizontal plane, 30° is a gentle dip, 60° is a steep dip, and a vertical bed dips 90°. A bed or a fault can be said to dip toward the southwest, for example. (Chap. 1)

Epoch A relatively short unit of geologic time, for example, Oligocene or Miocene. (Chap. 6)

Era A major unit of geologic time, for example, Mesozoic or Cenozoic. (Chap. 6)

Erosion The wearing away and removal of rocks at the Earth's surface by running water, waves, wind, or glaciers. (Chap. 2)

Eruption The pouring out of lava or the explosive release of ash from a volcano. (Prelude)

Evaporite A mineral that crystallizes from the highly saline brines that form when seawater evaporates. The most important evaporite minerals are halite (rock salt, or NaCl), anhydrite ($CaSO_4$), and gypsum ($CaSO_4 \cdot 2H_2O$). (Chap. 13)

Extension Horizontal stretching or pulling apart of the crust or sediments. (Chap. 11)

Fault A fractured break where rocks on one side of the break have slipped past the rocks on the other. (Chap. 5)

Faunal succession The observation, formulated as the "law of faunal succession," that species of fossils change in a definite sequence as one moves upward through a sequence of sedimentary rocks. (Chap. 1)

Flat A portion of a thrust fault that follows a weak zone parallel to bedding. (Chap. 11)

Foraminifera (*singular: Foraminifer*) Informally called "forams." Single-celled organisms that either live on the sea floor (benthic forams) or float near the ocean surface (planktic forams). Also the tiny shells constructed by these organisms. They are extremely important for dating sedimentary rocks. (Chap. 1)

Formation The basic unit of sedimentary rock, when classified on the basis of what kind of rock it is; e.g., limestone or sandstone. *Scaglia rossa* and *Majolica* are formations. This technical usage differs from the colloquial use of "rock formation" to mean some sort of oddly shaped erosional remnant. (Chap. 4)

Fossil The remains of an animal or plant that lived long ago, usually derived from a hard part like shell, bone, or wood, often replaced by mineral matter so that it is resistant to decay. (Chap. 5)

Fossil zone An interval of stratified sedimentary rocks characterized by the presence of a particular fossil genus or species, or a characteristic assemblage of fossils. Fossil zones are used for dating rocks on the basis of the animals or plants that were alive during their time of deposition. (Chap. 6)

Geomorphology The branch of geology that deals with understanding landscapes and the erosional processes that shape them. (Chap. 9)

Graded bed A bed of sandstone that has its coarsest grains at the bottom, with the grain size diminishing upward. (Chap. 10)

Granite A light-colored plutonic igneous rock whose coarse grains are mostly the minerals quartz and feldspar. Granite is abundant in continental crust and missing in oceanic crust. (Chap. 3)

Gypsum An evaporite mineral with the composition $CaSO_4 \cdot 2H_2O$. (Chap. 13)

Halite Rock salt (NaCl), a common evaporite mineral. (Chap. 13)

Hercynian Chain Mountains built in the late Paleozoic, about 300 to 350 million years ago, mostly in Europe. No longer actively deforming, they have been eroded off and now make up the continental crust on which the sedimentary rocks of the much younger Apennine Mountains were deposited. (Chap. 8)

Igneous rock A rock formed by the cooling and crystallization of molten magma. Igneous rocks are mostly either plutonic, meaning cooled at depth, so that there is time for large grains to form, or volcanic, meaning erupted at the Earth's surface, so that the quick cooling gives only very tiny grains. (Chap. 3)

Ignimbrite Another word for an ash-flow tuff. (Chap. 3)

Lahar A volcanic mudflow, formed when volcanic ash, soaked with water, flows down a slope. (Chap. 3)

Lava Molten rock that flows down the slope of a volcano before cooling. (Chap. 2)

Limestone A sedimentary rock made primarily through precipitation of the mineral calcite (or aragonite) by organisms, so that it has the same composition as calcite, $CaCO_3$. Most limestones form in shallow water, but pelagic limestones accumulate on the deep sea floor. (Chap. 1)

Listric fault A normal (i.e., extensional) fault that is curved, with the concave side upward. (Chap. 5)

Mantle The rocky portion of the Earth below the crust and above the metallic iron core, making up the largest volume of Earth material. Its top is between about 5 and about 40 km below the surface, depending on the thickness of crust, and its base is at a depth of about 2,900 km. (Chap. 2)

Marl A sedimentary rock with a mixed composition falling between limestone and clay. The clay component makes it softer and more easily eroded than limestone. (Chap. 7)

Marnoso-arenacea   The Formazione *marnoso-arenacea* (the marly-sandy formation) is a stratigraphic unit made of quartz-feldspar turbidite sandstones separated by gray marls, deposited in the Umbria-Marche Apennines during the Middle Miocene. (Chap. 10)

Metamorphic rock   A rock whose mineral composition has been altered, usually accompanied by deformation, because of heat and pressure during burial. (Chap. 3)

Microplate   A very small plate with dimensions of hundreds of kilometers at most, compared to the thousand-kilometer dimensions of the dozen or so main plates recognized in plate tectonics. Microplates are found mostly in tectonically complex regions like the Mediterranean. (Chap. 14)

Mineral   A solid grain with a particular chemical composition and a particular crystal structure, for example, quartz, feldspar, or calcite. Minerals are the building blocks from which nature constructs rocks. (Chap. 2)

Mudflow tuff   A rock that originated as a volcanic mudflow, formed when volcanic ash, soaked with water, flowed down a slope. (Chap. 2)

Normal fault   A fault whose motion has allowed part of the crust or its sedimentary cover to extend and stretch horizontally, usually causing younger rocks to drop down on top of older ones. (Chap. 12)

Oceanic crust   A layer of dense igneous rocks, mostly basalt, about 5 kilometers thick, resting on top of the mantle, underlying ocean basins. Oceanic crust is produced from the mantle by sea-floor spreading at mid-ocean ridges, and is recycled back into the mantle at subduction zones. (Chap. 6)

Outcrop   An exposure of bedrock, not covered by soil, vegetation, or any material emplaced by human activity. The common term for this is "outcropping," but geologists prefer to say "outcrop." (Prelude)

Paleogeography   The ancient configuration of lands and seas. One of the goals of geologists is to prepare accurate maps of paleogeography. (Chap. 8)

Paleomagnetism   The preservation in rocks of a record of the direction of the Earth's magnetic field at the time the rock solidified or was deposited. (Chap. 6)

Pelagic limestone   A pelagic limestone is made of tiny grains that settled slowly down through the water of the ocean onto the deep-sea floor. The grains are in part the shells of single-celled organisms that floated near the sea surface until they died and sank, and in part they are miniscule grains of clay that can drift long distances before setting to the bottom. "Pelagic" comes from a Greek word for the sea. (Chap. 6)

Period   An intermediate-length unit of geologic time, for example, Cretaceous or Paleogene. (Chap. 6)

Plate   One of approximately a dozen nearly rigid portions of the Earth's top 100 km or so. Each plate moves relative to the adjacent plates but does not deform internally to any significant extent. It might be better to call them "rigid spherical caps," which better conveys the geometry, but "plates" is universally used. Plates typically have lateral dimensions of thousands of kilometers, and most contain areas of both oceanic and continental crust. (Chap. 6)

Plate tectonics   The concept that tectonics—the growth and destruction of ocean crust and the building of mountains—is driven by the relative movements of large spherical caps, or plates, about 100 km thick, made of continental and oceanic crust and the uppermost part of the mantle. The theory of plate tectonics revolutionized geology in the 1960s and 1970s. (Chap. 6)

Plutonic rock   An igneous rock that cooled slowly at depth, giving time for coarse mineral grains to grow. (Chap. 6)

Ramp   A portion of a thrust fault that cuts diagonally upward through a strong interval of rock, joining two fault segments called flats, that follow weak layers in the bedding. (Chap. 11)

Ramp anticline   An anticlinal fold formed where rock layers are forced up over a thrust-fault ramp. (Chap. 11)

Rock   One of the natural inorganic materials that make up the Earth's crust and mantle. Rocks are aggregates of many grains of one or more minerals. (Chap. 2)

Scaglia   A kind of pelagic limestone common in the Umbria-Marche Apennines, dating from the Late Cretaceous and Paleogene. *Scaglia* means "flake" or "scale" in Italian, and refers to the way stonemasons can easily shape blocks of this rock by knocking off flakes with a hammer. (Chap. 1)

Sea-floor spreading   The process that takes place at mid-ocean ridges, in which mantle rises to the surface, partially melts, undergoes chemical transformation to basalt, and spreads out laterally in a symmetrical pattern on either side of the ridge, forming new ocean crust. (Chap. 6)

Sediment   Fine-grained, loose rock debris. Sediment may be formed through erosion, then transported by water, wind, or ice, and finally deposited to form new rock. Sediment may also be produced as the shells or shell fragments of animals, or by crystallization of salts when seawater evaporates. (Chap. 2)

Sedimentary rock   Rock that builds up through the accumulation of sediment. (Chap. 2)

Seismic   Referring to waves of vibration that pass through the Earth, generated either naturally, by earthquakes, or artificially, as part of a method for finding out what lies at depth below the Earth's surface. (Chap. 8)

Shale   A sedimentary rock that is composed mostly of clay minerals and thus is soft and easily eroded. (Chap. 10)

Stage   A very short unit of geologic time, for example, Maastrichtian or Messinian. (Chap. 6)

Subduction   The process by which old ocean crust sinks back into the mantle, balancing the formation of new ocean crust at mid-ocean ridges. (Chap. 9)

Superposition   The law of superposition was Nicolaus Steno's great discovery, in the 1660s, that history is recorded as growth bands in solids, and most important, that younger sedimentary rocks rest on top of older ones. (Chap. 5)

Syncline   A fold where beds close downward, making a structural trough. (Chap. 7)

Tectonics   Study of the Earth's largest-scale geologic structures—continents, ocean basins, mountain ranges, plateaus, and sediment-filled basins. (Chap. 7)

Thrust fault   A fault whose motion has allowed part of the crust or its sedimentary cover to shorten horizontally, usually driving older rocks up over younger ones. (Chap. 8)

Transform fault   A plate boundary along which two plates slide past each other, neither generating new crust as at a spreading ridge, nor consuming old crust as at a subduction zone. (Chap. 9)

Trench A very deep, elongated part of the sea floor, marking the site where oceanic crust is being subducted. (Chap. 9)

Tuff A volcanic rock made of volcanic ash and larger volcanic fragments, which may accumulate in a variety of ways (see air-fall tuff, ash-flow tuff, welded tuff, ignimbrite, lahar). A tuff is thus a sedimentary rock, although the sediment owes its origin to volcanic processes. (Chap. 3)

Turbidite A sandstone bed, typically graded from coarse at the base to fine at the top, deposited by a turbidity flow. (Chap. 10)

Turbidity flow A mixture of water and sediment, commonly generated when an earthquake shakes the seabed. A turbidity flow sweeps down the slope of the sea floor or of a lake bottom because the bulk density of sediment plus water is greater than the density of water. (Chap. 10)

Unconformity A break in the stratigraphic record, where there are no rocks to represent some of the time that passed. (Chap. 8)

Volcanic ash Very fine particles of volcanic rock fragmented during an explosive eruption. This is rock matter, not to be confused with the ash left after wood burns. (Chap. 3)

Volcanic rock An igneous rock erupted from a volcano, either as molten lava or as finely fragmented volcanic ash. Because of the limited time for cooling, the mineral grains in volcanic rocks are generally very small. (Chap. 1)

Welded tuff An ignimbrite or ash-flow tuff carried by gases so hot that after the flow came to rest, the ash was softened by the heat and became welded together into a very hard rock. (Chap. 3)

ACKNOWLEDGMENTS

This book owes so much to so many people. Rick Dehmel first convinced me to write it, and his elegant sense of storytelling helped me again and again. Jack Repcheck was not just the editor but provided the friendship and inspiration that kept the project on track. I have benefited from many valuable suggestions from Massimiliano Barchi, Daniel Bernoulli, Alberto Castellarin, David Christian, Lisa Lamb, Bill Lowrie, Chad Manning, Eldridge Moores, Kevin Padian, David Shimabukuro, and Enrico Tavarnelli. Sara Pertusati's viewpoint as one of the new generation of Italian geologists kept my older recollections in perspective. Charlie Bond's careful reading of the manuscript helped me see the story as a meditation on time, and Joe Vinikow contributed a clear vision of the dramatic character of the challenge to understand Earth history. I am grateful to Arnold Bouma for recollections of the early days of turbidite sedimentology, and to Elio Ciol, Howie Spero, Ferruccio Farsi, and Stefan Schmid for photographs. The push to finish came from a group of Berkeley students—Kealan Cunningham, Sarah Dayley, Courtney Guerrero, John P. Hughan, Laurel Lemontt, Sonia Malek, Christian Placencia, and Ziwei Victoria Tang—whose feedback also helped to hone each chapter.

    I have had the pleasure of knowing a great many Italian geologists, beginning in the 1960s—some for a single conversation, some for field trips or extended research projects, and some who became lifelong friends. The Italian geologists have built our understanding of the Alps and Apennines, and they are the heroes of this story. When I started compiling a list that contained almost a hundred names before I even got to the end of the 1970s, I realized that it was just not feasible to mention every Italian geologist I have met. Still, I warmly thank them all for their many kindnesses to a fellow geologist from another country.

Over the years I have been blessed with wonderful students, postdoctoral fellows, professional colleagues, and scientific visitors, whose contributions to our work on Italian geology have made this book possible: Luis Alvarez, Albert Ammerman, Mark Anders, Mike Arthur, Frank Asaro, Massimiliano Barchi, David Bice, Fausto Burattini, Stefano Cardellini, Alberto Castellarin, Michael Carr, Ernesto Centamore, Lung Sang Chan, Jim Channell, Philippe Claeys, Aron Clymer, Tommaso Cocozza, Roberto Colacicchi, Ben Crosby, Garniss Curtis, Ian Dalziel, Bruno D'Argenio, Brett Davidheiser-Kroll, Tonino Decandia, Al Deino, Don DePaolo, Bob Drake, Terry Engelder, Fernando Falco, Joshua Feinberg, Al Fischer, Bruce Fouke, Joan Gabelman, Peter Geiser, Andy Gordon, Manuel Grajales, Mark Heckman, David Jones, Dan Karner, Peter Keller, Roy Kligfield, Christian Koeberl, Paul Kopsick, Victoria Langenheim, Giusy Lavecchia, Bill Leith, Jonathan Levine, Carolina Lithgow-Bertelloni, Bill Lowrie, Stan Margolis, Steve Marshak, Paolo Mattias, Peter Mattson, Mike McWilliams, Annemarie Meike, Helen Michel, Giorgio Minelli, Alessandro Montanari, Sylvia Moses, Rich Muller, Giovanni Napoleone, Diane O'Connor, Sara Pertusati, Giampaolo Pialli, Chuck Pillmore, Isabella Premoli Silva, Ed Rashak, Paul Renne, Domenico Rinaldini, Bill Roggenthen, Mark Rowan, Ettore Sannipoli, Rich Schweickert, David Shimabukuro, Gene and Carolyn Shoemaker, Jan Smit, Erick Staley, Kevin Stewart, Nicola Swinburne, Enrico Tavarnelli, Nicola Terrenato, Brent Turrin, Luigi Vigliotti, John Wakabayashi, Christine Waljeski, Chi-Yuen Wang, Lionel Weiss, Rudy Wenk, and Forese Wezel.

Most of this book was written in Berkeley, with parts composed at the Osservatorio Geologico di Coldigioco in Italy and at the Helen Riaboff Whiteley Center of the Friday Harbor Laboratories, in the San Juan Islands of Washington.

Through all the years of fieldwork in Italy, Milly Alvarez—taking well-earned breaks from her work in mental health—has been the best of companions, relishing not only the interesting people and the beautiful scenery but also the most uncomfortable quarters and the most difficult hikes through boulder fields, deep mud, and thornbushes, and up steep mountainsides in the blazing sun, delighting in good weather and bad, in new places and in old friends.

ADDITIONAL READING

For readers of this book who want to follow up by reading more about geology, here are some suggestions:

Pompeii: A Novel, by Robert Harris (Random House, 2003), is a fine historical novel about the eruption of Vesuvius that destroyed Pompeii and Herculaneum in A.D. 79, full of information about volcanic geology and Roman civilization.

John McPhee's stories about geology and geologists, collected as *Annals of the Former World* (Farrar, Straus & Giroux, 1998) have delighted and instructed a wide audience over two decades.

Richard Fortey takes readers on a global geological tour in *Earth: An Intimate History* (Alfred A. Knopf, 2004).

Very readable accounts of three pioneering geologists have appeared recently. For the life of Nicolaus Steno, see Alan Cutler's *The Seashell on the Mountaintop* (Dutton, 2003). James Hutton is recalled by Jack Repcheck in *The Man Who Found Time* (Perseus, 2003). Simon Winchester tells the story of William Smith in *The Map That Changed the World* (HarperCollins, 2002).

E. B. Bailey's *Tectonic Essays, Mainly Alpine* (Oxford University Press, 1935 and 1968) is a delightful story of the early geologists in the Alps. For the more recent history of geology, Naomi Oreskes has written two accounts of the change from fixed continents to plate tectonics: *The Rejection of Continental Drift* (Oxford University, 1999) and *Plate Tectonics: an Insider's History of the Modern Theory of the Earth* (Westview, 2001).

Chris Scholz conveys the adventurous mystique of field geology in *Fieldwork* (Princeton University Press, 1997).

For those interested in a systematic introduction to geology, my favorite textbook is *Earth: Portrait of a Planet*, by Steve Marshak (W. W. Norton, 3rd ed., 2008). A fine text with a planetary perspective is *Earth: Evolution of a Habitable World*, by Jonathan Lunine (Cambridge University Press, 1998). A compact account of how Earth came to be is given by Wally Broecker, *How to Build a Habitable Planet* (Eldigio Press, 1985).

Ken Deffeyes's *Hubbert's Peak* (Princeton University Press, 2001) is the best introduction I know to the geology of petroleum, with a focus on the question of how much is left.

The literary aspects of geology are explored in *Bedrock: Writers on the Wonders of Geology*, edited by Lauret Savoy, Eldridge Moores, and Judy Moores (Trinity University Press, 2006).

Finally, my own book *T. rex and the Crater of Doom* (Princeton University Press, 1997) tells in more detail the story of impact and dinosaur extinction, based on discoveries in the Mountains of Saint Francis.

INDEX

Page numbers in *italics* refer to illustrations.